NATHANAEL-ISRAEL ISRAEL, PhD

Reconciling Science and Creation Accurately

OTHER BOOKS BY NATHANAEL-ISRAEL ISRAEL

Get them at your local bookstore, or online (e.g. on Amazon, Science180.com/books)

Turbulent Origin of the Universe
There is Only One Scientific, Simple, Safe, Trustworthy, Unexpensive, Brave, Practical, Nonconformist, Universal, Verifiable Formula that Accurately Decodes the Universe Formation … But You Are Not Using It

Turbulent Origin of Chemical Particles
Why You Don't Have to Embrace Evolution, Big Bang, or Deny God to Scientifically Prove the Formation of All Chemical Particles

Origin of the Spiritual World
Top Secrets about the Origin of Everything in the Universe that Some Elites Have Hidden from You for Thousands of Years

From Science to Bible's Conclusions
How Decoding the Universe-Origin by Properly Revisiting Scientific Data—That Top Scientists Collected but Wrongly Analyzed—Bizarrely led to the 3500 Years Old Biblical Account of Creation

Turbulent Origin of Life
Why You Don't Have to Embrace Evolutionism or Check Your Brain at the Door in the Name of Faith or Science to Accurately Decrypt the Origin of Life Using the Historic Formula of the Universe Formation

How God Created Baby Universe
What Science Accurately Teaches about Creation and God's Existence that Atheists, Freethinkers, and even Most Christians Ignore … And How to Demonstrate it Without Taking Sides Between Rationality and Faith

How Baby Universe Was Born
How to Scientifically Talk to Children about the Universe Formation and They will Know Forever How to Correctly Test the Intersection of Science and Faith

Science180 Accurate Scientific Proof of God
Can We Scientifically Explain the Formation of the Universe Through Natural Processes Without Evoking Evolution and Big Bang?

Mathematical Proof of God's Existence at the Intersection of Science and Faith.
The Scientifically Verifiable Cosmological Theory that Challenges the Big Bang Theory at the Crossroads of Reason and Religion THEY Want You to Ignore

More books written by Nathanael-Israel Israel can be found at Israel120.com/books

NATHANAEL-ISRAEL ISRAEL, PhD
Founder of Science180: www.Science180.com
Father of Science180 Creationism.
Discoverer of the Turbulent Universe-origin Formula
www.Israel120.com

Reconciling Science and Creation Accurately

What Science Accurately Teaches about Creation and God's Existence that Atheists, Freethinkers, and even Most Christians Ignore …
And How to Demonstrate it Without Taking Sides Between Rationality and Faith

$$T = \frac{D}{Ve} + \frac{2\,R\,\pi}{Vo}$$

Science180

Augusta
United States of America
www.Science180Publishing.com

Reconciling Science and Creation Accurately
What Science Accurately Teaches about Creation and God's Existence that Atheists,
Freethinkers, and even Most Christians Ignore ... And How to Demonstrate it Without
Taking Sides Between Rationality and Faith

First edition: October 2025

Published by Science180
Augusta (USA)
www.Science180Publishing.com

Book Cover and Illustrations by Nathanael-Israel Israel

ISBN: 979-8-9932150-2-0

Library of Congress Control Number: 2025920946

Printed in the United States of America.

CONTENT

DISCLAIMERS

Above all, readers of my writings must understand that my goal is NOT, and will NEVER be, to go against God and His message in the Bible. No one should ever consider my interpretation of the Bible and the scientific data available at the time of this book's publication to contradict the Word of God revealed in the Bible. My prayer is to positively impact many lives, revolutionize science in a better direction, impart true knowledge to people, strengthen believers' faith, and help many unbelievers come to Jesus. Although I believe that God revealed certain things to me, I am also conscious that, as a writing by a human being, this book, and others that I wrote, may contain mistakes, which may be found and addressed later as more revelations are known and more scientific data are discovered or better analyzed. Therefore, no one should ever consider my commentaries as absolute and/or free from errors. I did my best to convey that I consider the finest interpretation of the scientific data and revelations known as of 2025, the year I published this book. Should any error still exist in this book and the others I wrote, I ask God to forgive me, and help me and my readers to detect and fix them for His glory. For to God alone be the glory, the honor, and the praise for creating such a perfect world, and for giving us the grace to live in it and prepare ourselves for eternity. I also pray that this book helps you grow in your understanding of the biblical creative narrative and your journey to heaven. When we get to heaven (I hope you believe in God and will make it there), we will definitely know more about creation. Until I have a chance to hear from you or talk with you about this book, be careful and prayerful about what you think you know and what you are about to read. May the knowledge of the glory of God fill the Earth as the waters cover the sea.

March 5th, 2020 (USA), i.e., 5 years 7 months before the publication of this book!
Nathanael-Israel Israel, PhD
www.Israel120.com

RECONCILING SCIENCE AND CREATION ACCURATELY

CHAPTER 1

HOW TO SCIENTIFICALLY INTRODUCE THE UNIVERSE'S FORMATION AND HAVE NONBELIEVERS, AND ALL OTHER FREETHINKERS, RATIONALLY BOW TO THE BIBLE?

1.1. Purpose of this book

Why am I teaching the whole world how science accurately supports the biblical account of creation? Indeed, the search for the origin of the universe is one of the ancient questions that humankind has been asking since antiquity:

- Where do we come from?
- How did the universe come into existence?
- Who created the universe?
- What are the forces or laws that govern the universe?
- How can science explain the origin of the world?
- Why are we here?
- Where are we going?

By a confluence of questions after my studies in the USA, where I earned a PhD (with distinction) in Plant, Insect, and Microbial Sciences. During my research on the origin of the universe, I realized that, from unbelievers to believers, many errors were made in answering those questions or in interpreting the creation stories, including those revealed in the Bible. For instance, the Genesis 1 story, credited to Moses in the Bible, has been the foundational account for Jews and Christians who believe that God created the world, although they do not always agree on the duration and process of creation. For instance, because they do not understand the process, speed, and power with which God created the world, some people (even believers) think that the creation days were not 24 hours each. Using some biblical verses, some people even think that a creation day can refer to thousands, millions, or even billions of years.

Religions around the world seem to interpret the creation story differently, and most theories about the origin of the universe do not fully take advantage of the depth of information hidden in the Bible. Those who do not believe in God have forged their own cosmology. Many believers who tried to incorporate the Genesis 1 narrative into their theories failed to properly address the scientific data. One of the challenges is that the book of Genesis was written before the scientific era, in a language most modern scientists failed to understand or recognize using simply current scientific terminologies. Another problem is that the universe is so vast and its constituents so diverse that it is hard for human beings to comprehend its formation and functioning solely through physical means.

In 2013, I was led into deep thoughts about how the world began, how it functions, and what I can do to succeed in life despite all the problems it has thrown at me. Toward that end, I have spent more than 7 years studying the bodies in the universe and unearthing laws that support their origin. I ended up studying more than 500 variables related to the celestial bodies (e.g. galaxies and their clusters, stars and their clusters, planets, planetary systems, asteroids, asteroid systems, satellites, satellite systems, and rings), and more than 100 variables related to microscopic particles (e.g. subatomic particles and atoms), macroscopic particles (e.g. minerals and rocks); I also studied a lot of things invisible to the naked human eyes (e.g. spirits).

Before I started this holistic study, certain things were made known to me at a time when I was not seeking them, meaning a time when I never even imagined myself thinking about such things. As time went by and I started investigating (using a secular research methodology), I discovered an astonishing match between what I had received years ago, particularly in 2013, and what I was finding in the literature. This observation made me take the research I was doing very seriously, for I realized I was a man on a mission. The details of the demonstration were not fully disclosed to me. Still, I knew I needed to find them using a creative investigative method that differed from conventional scientific research. Part of my mission was to prove what I was shown was what I found, using means that even the stubborn unbelievers must accept, provided they are willing to follow the facts wherever they lead. In the process, I realized that I was not called for the church where I would list one biblical verse after another to try to prove how God created the world, for such ways of reaching out to unbelievers has not been working and will not work with most educated people, who are so determined to reject anything attached to God unless it is physically proven. I realized that I was called to open the eyes of the most tenacious unbelievers using some of the means they use to do their own research, show them the scientific thought process I used to find what I found (although everything cannot be found using secular scientific methods that is not after the truth), so that in the process, their minds can be inclined to accept the conclusions, which most conventional preachers present to them in a language that those unbelievers will never accept. Furthermore, although they are in a big problem of unbelief, some unbelievers seem not to have some apparent critical physical needs or apparent common problems (e.g. sickness, hunger, thirst for knowledge, questioning about suffering, outcome of all possible options for the solutions of their problems) that usually dispose some stubborn people to listen to those who approach them with the word of God; for those so-called smart or rich people think that they can use their money to solve all their problems (which some of them would even deny), or they do not want to make room to hear about the possibility of a divine being above their power, influence, wealth, etc.

Therefore, to accommodate those who don't believe in God, I had to be very careful so that my efforts to minister to them, or at least to give them a chance to read the findings of my investigation, would not be sabotaged by the number of undemonstrated facts I would start sharing with them. Some of these stubborn unbelievers have already been exposed to the truth and facts by people who wrapped or packaged them in evangelization envelopes they do not like or they do not even want to approach; for their definition of knowledge, prosperity, areas of interest, things worth spending time on, priorities, and other pursuits of most so-called educated or rich people do not usually include religious debates or dogma that unfortunately some religious people fail to properly demonstrate (not because they do not want to, but because the truth and meanings behind many things and reality in the universe is highly encrypted in messages, which cannot be decrypted if the one wanting to decrypt them already refused the conclusion or content of that message, and also that decryption needs an amount of time greater than the common lifespan of a mere human being); for the secret of the password of the encrypted message in the universe, which some scientists are looking for, can never be fully physically known by scientific method only, but at the end of time or by people who can travel through time (past, present, and future) to see what it would be and what was said about it.

Therefore, even after I discovered the codes of the universe formation and functioning, I realized that it is impossible to address and give them a chance to be read first by people before they reject them—still by sticking to the principle of the means required for anyone to find them—if I do not present the story in a form that allows people to always have a way to reject them before reading them and also to accept them before finishing reading them. This challenge is true not only for unbelievers but also for believers. Hence, I wrote a completely separate book just for the scientists, but in a way that can draw those who are thirsty or hungry for the truth to seek more and to read the religious details covered in this book. However, when I started addressing the religious aspects of my findings about the formation of the universe, I also realized that the believers do not have the same understanding of the scriptures, and some of them would not even want to read the book if it had some references about authentic books (e.g., the Book of Enoch, the Book of Jubilees, and the Book of Jasher quoted in the Bible) that are not in the

Bible—because some of the people who compiled the Bible were uninspired and rejected them, not because all of those books are false, but because some are too deep to be easily understood by some people, as mentioned. Even in this version of the book (which I called the biblical version, or the prophetic version), there are other biblical facts that I could not mention because they would be rejected by others who may still refuse to acknowledge that some rejected books or lost books of the Bible are worthy to be read and weighed. Moreover, because Bible believers have various interpretations of Bible verses, I had to remove certain chapters (I originally wrote), but that may appear controversial to others because they do not believe, yet they are true! In other words, to try to accommodate as many people as I could without diluting my message, on top of *"Reconciling Science and Creation Accurately"*, I also wrote another book, *"Origin of the Spiritual World"*, which I call the "supra prophetic version" or the "pseudepigraphic version,", in which I dealt with things of a higher level of belief, revelation, or prophecy related to the universe. In the supra-prophetic version, I addressed topics such as the story of creation written by Enoch, the man who (the Bible said) pleased God so much that God took him to heaven alive! Therefore, I did two versions of the religious book:

- the Biblical or prophetic version: the current book you are reading (which, just because of its label—prophetic— some people who deny the prophetic may not even read, but I had to call it the prophetic version, for without the prophetic insight, it is impossible for me to write it) and

- the supra prophetic version (that I also called the pseudepigraphic version), which I think is of a higher level of the prophetic, for in that version, I addressed supernatural things that only those who are deep in the prophetic—and those who have done some homework regarding life, the word of God, the origin of the universe, etc., beyond the mere teaching in science and some theological schools—can accept.

To better appeal to different audiences and ensure they read the book, I had to choose what to share accordingly. By the time I published this book in 2025, I had also released 8 others, including 2 targeting children and one solely addressing the scientific proofs of God's existence. In other words, to fully grasp all my findings about the formation of the universe, it would require people to read all the versions I wrote on that topic:

- *"Turbulent Origin of the Universe"*: The scientific version (labeled as such not because the other versions of this book do not contain scientific facts, but because this version contains the bulk of my scientific demonstrations; yet, for the sake of space, I had to cut the size of that book by more than half)

- *"From Science to Bible's Conclusions"*: The public version (a summary of all my books on the origin of the universe)

- *"Turbulent Origin of Chemical Particles"*: The chemical version (related to the origin of chemical particles)

- *"Reconciling Science and Creation Accurately"*: The Biblical version (the current book you are reading)

- *"Origin of the Spiritual World":* The pseudepigraphic version (prophetically more advanced than the current one you are reading), and
- *"Turbulent Origin of Life":* The biological version (which focuses on the origin of life)
- *"How Baby Universe Was Born": a* children's version targeting unbelieving children and their parents
- *"How God Created Baby Universe":* a children's version targeting believing children and their parents
- *"Science180 Accurate Scientific Proof of God":* written anyone seeking the real scientific proofs of God's existence

For instance, in my book *"Turbulent Origin of the Universe,"* I could not address certain facts that some unbelievers might quickly attack (because they contain biblical references). In other words, although I found myself writing several books to explain the origin of the universe, I was not comfortable combining all my thoughts into one book, for it would be too complex, too simple, or too diluted, and most people would not understand it. If I had tried to put everything into a single book, many people would not read it (because it would be too heavy and very complex), and my efforts could have been wasted and the reader's expectations deceived. Similarly, to avoid diluting the message for the sake of just pleasing people (or to make everybody understand everything), and to avoid chasing people away because I presented the conclusion; regardless of the version of the books I wrote, I chose to tell a story mixed with the facts I investigated without always stating the logic behind the steps I used, but by the time the readers go over all of the book(s), they should understand the reasons behind what I did and choose for themselves, without thinking that my goal is to force them into accepting something they do not want or that they are predisposed to never accept. Hence, although I wrote the first version of the scientific and religious books in 2020 (in the middle of the coronavirus pandemic), it took me 5 more years to break them down into separate books, which can relate to different markets or people, according to their scientific, religious background, or interests.

Indeed, there is only one simple, compelling, solution-directed scientific formula that is accurate enough to explain rationally how God created the universe. *"Reconciling Science and Creation Accurately"* is a landmark book in universe-origin writing from a rare perspective by one of the most respected minds of our time. It scientifically explores the most challenging questions of all time that interest believers, nonbelievers, and all freethinkers: How can we rationally demonstrate, without checking our brains at the door in the name of faith, that God created the universe? How did the universe begin, and what processes did God use to create it? Are these processes still operating in the universe or not? Can believers abandon wrong theories if they think it is impossible for science to prove the Genesis story literally, or if they think that science is evil and opposed to faith, or if they think they must compromise by embracing scientific theories that contradict the biblical account of creation written before the scientific era? What can believers do to help the skeptics believe in the biblical narrative of creation?

Lucky you, Dr. Nathanael-Israel Israel successfully navigated all those questions with an accuracy that has been applauded by both scientists and nonscientists across the globe. After reading *"Reconciling Science and Creation Accurately,"* you will confidently:

- Scientifically prove the Biblical account of the creation of the universe and the existence of God in a way that makes the heads of those who deny God spin faster than a DJ's turntable
- Know how to rationally talk to anti-creationists, evolutionists, Big Bang proponents, atheists, skeptics, and other freethinkers about the universe's formation, and they will beg you to know more about God, the Creator, whom they mistakenly rejected
- Discover very accurate, rare, and factual truths about the universe's origin that will save you time and money, and get you much closer to the better and joyful life you want to live today and forever
- Improve your health and faith by knowing that the existence of God can be scientifically justified using Science180 Cosmology and particularly Science180 Creationism
- Enter a new area of freedom and power by crushing the head of and breaking free from the suffocating expectations of all wrong theories that have hijacked secular and religious education, and that have held the Biblical account of creation captive for almost 3500 years
- Break free from the suffocating expectations of some forms of creationism that have sequestered the minds of some believers for a long time
- Uncompromisingly, intelligently, and scientifically explode the myth of those who, instead of literally taking the Biblical days of creation as 24-hour consecutive days, think that they were millions of years, or were representative of long ages, or that millions of years existed before them or were positioned between them
- Understand the accurate standard to interpret the Biblical account of creation, thanks to Science180's breakthrough that transformed science and laid a foundational bedrock for the inerrancy of Scripture

Now that Genesis (the oldest manuscript in the world, written before science and most religions were born) is scientifically proven to be correct (*Science180.com*/biblical), what unstoppable, jaw-dropping paradigm shift will the discovery of the perfect alignment between science and the Bible bring into the religious, rational, and secular world today? Keep reading this thoughtful book to figure out what happened at the beginning, what is coming up, and why it is time to urgently rethink everything you have been told about the universe's origin so you

CHAPTER 1: INTRODUCTION

don't eventually regret it! Don't say that nobody told you!

Founder of Science180 Academy, Dr. Nathanael-Israel Israel (the author of this book) is acknowledged worldwide as the discoverer of the all-in-one, proven, and simple scientific formula that accurately cracked the origin of the universe, of life, and of chemicals, and that scientifically unearthed the holy grail at the intersection of science and the Biblical account of creation. Learn more at Israel120.com.

By the time you finish reading this mind-blowing masterpiece, you will:

- Add an efficient, trustworthy, and cost-effective program that you truly need on your strategic journey toward your best tomorrow
- Become a transformational catalyst that knows how to rock the boat of the organizations that are engaging in activities that deny God, the Creator of the universe
- Break free from the suffocating expectations of all wrong theories (e.g., Evolutionism, Big Bang, and even some forms of creationism) that have hijacked secular and religious education for so long
- Break free from the suffocating expectations of some forms of creationism that have sequestered the minds of some believers for a long time
- Discover how the accurate reanalysis of the scientific raw data from a fresh perspective bizarrely landed on the conclusion of the Bible, clarified contradictory interpretations of the Biblical account of creation, and reconciled science and Genesis while rationally uncovering key factors that have influenced the relationship between science and faith
- Discover the all-in-one proven and uncomplicated formula to accurately reconcile science and the Bible
- Discover the scientific formula that proves the existence of God
- Discover the scientific formula that successfully tested the relationship between science and the Biblical account of creation
- Educate yourself to better resonate with your target market or customers who are craving something original that break anti-Biblical explanations of the origin of the universe and life
- Empower yourself to leave unforgettable marks in your field of expertise
- Enjoy multiple origin-related programs and choose the ones that suit your needs
- Enter a new area of freedom and power that crush the head of Evolutionism, Big Bang, and all other anti-creationist theories that have held the Biblical account of creation captive for so long
- Know everything all believers have ever wanted to know about the creation of the universe and its contents
- Fearlessly push the boundaries of human abilities to properly understand what is perceived as un-understandable, mysterious, supernatural, unimaginable, impossible, and unthinkable that hold you back

Science180: A Simple Universe-Creation Theory That Made No Assumption

- Free yourself from the asphyxiating expectations of all scientific theories of the universe's formation that have sequestered the minds of some believers for a long time
- Get step-by-step instructions about how to properly demonstrate the creation of the world
- Harness proven universe-origin tools to holistically pave your way to higher ground and properly reject all pressures to scientifically or philosophically fit in with anti-creationist models
- Hear Dr. Nathanael-Israel Israel's personal selection and teaching of key topics that will help you break the code of the universe's formation and functioning
- Interact with a renowned expert who will not just lecture you, but help you sort out your origin-related questions using strategies to tap into deep secrets you ignore
- Learn and teach the scientific proofs of the existence of God in a way that will make the heads of those who deny God spin faster than a DJ's turntable
- Learn from an easy-to-work-with expert who will respond to your universe-formation needs and position you to stay on top of your competitors
- Learn from the historic specialist of universe-origin questions: what is the proven formula that demonstrates the formation of the Earth on the 3rd day of creation, and the formation of the Moon and Sun on the 4th day, just as the Bible revealed
- Learn how Science180 turned all existing cosmological theories and all incorrect established norms and absurdities of the universe's origin on their head by accurately reconciling Science and the Biblical account of creation
- Learn how to scientifically demonstrate that the days in the Biblical account of creation were literally 24-hour consecutive days, a milestone that accurately reconciled science and the Bible, and that overturned a myth according to which some people have thought that each day of creation was millions or billions of years (a misunderstanding that caused many people to deny God, the Creator)
- Learn how to scientifically unlock and interpret the Biblical account of creation without checking your rational brain at the door
- Learn key life lessons to improve lives instead of repeating mistakes that many people still ignore at their own peril.
- Listen to an experienced expert who discovered outstanding secrets about the origin of all there is
- Logically, learn how the Sun was formed in less than 4 days
- Participate in the global effort that is lighting an unquenchable fire under all traditional and scientific nonsense about the origin of the cosmos

- Scientifically defy all existing theories that rob God of his glory in creating the universe and its content
- Scientifically learn how the Earth was formed in less than 3 days
- Sharpen and optimize your creation-related strategies so they can effectively convince and convert the unbelievers by using scientific facts backing the Biblical account of creation
- Stand tall as a symbol of freedom, power, creativity, and originality in your field of expertise
- Systematically learn how the Moon was formed in less than 4 days
- Understand the accurate standard to interpret the biblical account of creation thanks to a breakthrough that transformed science and laid a foundational bedrock for the inerrancy of Scripture.

1.2. Keys things to consider to scientifically explain the universe-origin

Have you ever wondered, "How can I scientifically understand the Biblical creation quickly without falling into the trap of taking sides between science and faith? Have you ever struggled with the interpretation of some verses in the Genesis story of creation, not knowing what to do or what sense to make of them without doubting God or checking your brain at the door? Well, I want to share with you a groundbreaking approach to accurately explore the relationship between science and the Bible and to answer critical questions surrounding the book of Genesis, the most-attacked book of the Bible.

The universe is filled with diverse bodies and systems of bodies, which, for the sake of space and time, I could not present here, but which I extensively reviewed in the electronic version of this book and others I wrote on the origin and formation of the universe (www.Science180.com). Everything in the universe can be classified into 2 groups: the physical world and the spiritual world, or the invisible world. Examples of things belonging to the spiritual world include God, angels, human souls, and human spirits. Here, I defined the physical world as consisting of things that can be seen by the naked eye or with scientific tools, such as microscopes or telescopes. Matters in the universe are hierarchically clustered into structures ranging from subatomic particles (meaning smaller than or located within atoms) to the largest scales, such as that of the highest galaxy clusters. Some subatomic particles can be clustered into atoms, which in turn can be clustered into molecules and chemical compounds, which in turn can be clustered into minerals and rocks, which are parts of celestial bodies.

Celestial bodies can be classified as planets, asteroids, satellites, stars, and the clusters of bodies they form, such as planetary systems, stellar systems, galaxies, and galaxy clusters. A stellar system contains at least a star. Some stars are isolated, while others are arranged into globular clusters and others into galaxies. In their turn, galaxies can be organized into galaxy clusters, some of which are arranged into galaxy superclusters, which can be organized into large-scale galactic filaments, and so on and so forth until the largest structure of the universe is reached. Apostle Paul, for instance, spoke of visiting the 3rd heaven, while other Jewish books spoke of 10 heavens. Considering the Jewish and Biblical classification of heavens, including biblical verses that said that God rides on the heavens, meaning that He does not live in heaven but on the heavens, there are many clustering levels of celestial bodies if heaven can be considered as a kind of cluster of bodies. Many contemporary prophets have also testified to having visited heaven. In *"Origin of the Spiritual World,"* I elaborated more on these mysteries.

Galaxies are known to contain thousands or even millions of stars. Just like the Sun, most stars are believed to have their own stellar system consisting of a star orbited by planetary systems and asteroids. Some asteroids also have their own systems, meaning they are orbited by a satellite. In general, planetary systems consist of a primary planet orbited by a satellite (if any) or a system of satellites (if more than 1). Although satellites are generally known as bodies orbiting planets, it is important to notice that some asteroids also have their own satellites. In *"Turbulent Origin of the Universe,"* the scientific version of this book, I deepened the characteristics of each type of celestial body and its particles. The planets, asteroids, and satellites I studied in depth are in the Solar System, which belongs to the Milky Way Galaxy. In fact, I studied more than 461 celestial bodies in the Solar System, including the Sun and 460 bodies orbiting it: 9 planets, 241 asteroids, 210 satellites, ring systems, and the atmosphere and crust of some celestial bodies. The names of the planets are Mercury, Venus, Earth, Mars, Jupiter, Saturn, Uranus, Neptune, and Pluto (although since 2005, some people no longer consider Pluto a planet). While some of these planets (Mercury, Venus, Earth, Mars, and Pluto) are terrestrial, meaning they have a hard crust, others are giant gas planets (e.g., Jupiter and Saturn), while others are giant ice planets (e.g., Uranus and Neptune). Fig. 1 illustrates the organization of the celestial bodies in the Solar System.

The 210 satellites known in the Solar System as of 2021 are the Moon (Earth's satellite), 2 Martian satellites, 79 Jovian satellites, 82 Saturnian satellites, 27 Uranian satellites, 14 Neptunian satellites, and 5 Plutonian satellites. At the publication of this book in 2025, more satellites were discovered last year, and the number will keep growing until all the satellites in the Solar System are discovered. Then, the debates about whether all of them are worth calling a satellite will start (if not yet).

CHAPTER 1: INTRODUCTION

Fig 1: Sketch of the planets and main belt asteroids in the Solar System

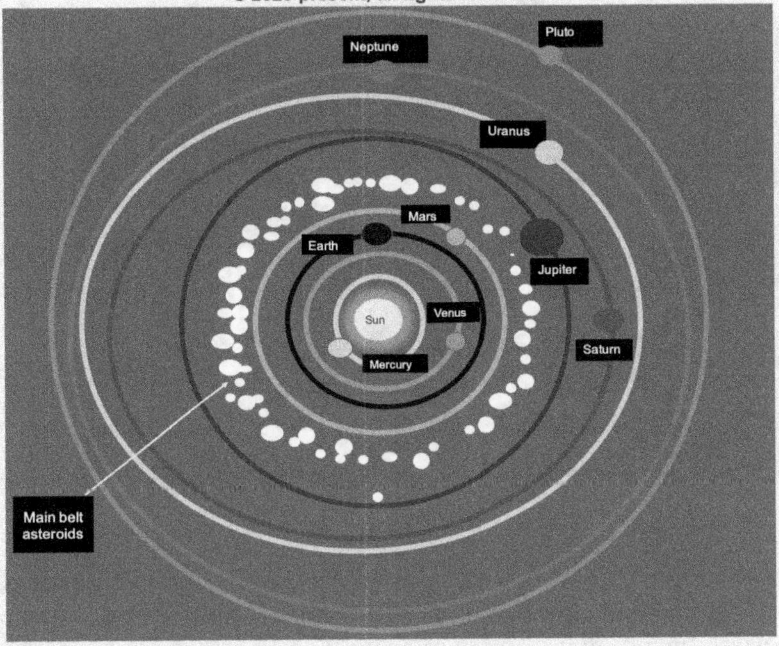

The Moon is the most obvious body in the night sky. With time, many other satellites were discovered, beginning with the work of Galileo, who opened the door to the exploration of the satellites around planets. Indeed, using his newly built telescope, Galileo discovered the 4 largest Jovian satellites in 1610. Afterwards, it took 45 more years before Titan, the biggest Saturnian satellite, was discovered. By the end of the 19th century, only 21 satellites had been discovered in the Solar System. In the first 20 years of the 21st century, only 140 more satellites were discovered in the Solar System, bringing the total number of satellites discovered since the time of Galileo to 210. As of 2021, no satellite has been found around Mercury or Venus. Yet new satellites continue to be discovered around other planets in the Solar System, but most are smaller, usually less than 2 km in radius. For instance, when I began working on the bodies in the Solar System in 2013, only 178 natural satellites were known to orbit planets. However, 32 new satellites in the Solar System were discovered from 2016 to 2019, prompting me to rewrite certain chapters on satellites. In 2025, more than hundred were added to the existing list.

Science180: A Simple Universe-Creation Theory That Made No Assumption

Fig. 2: Sketch of the orbits of satellites around a primary planet

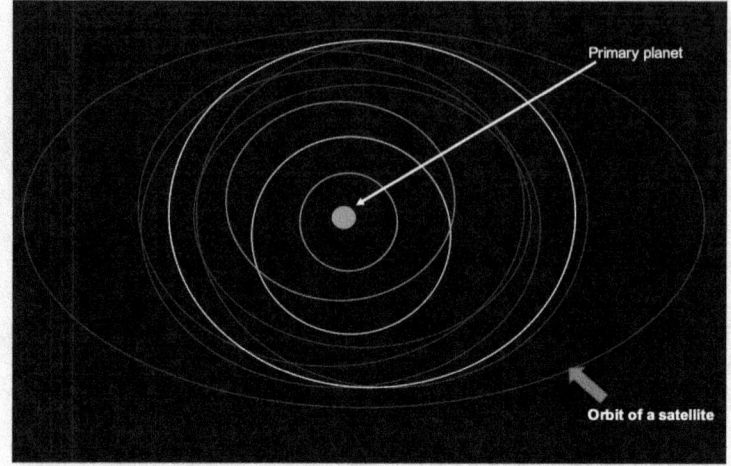

The satellites in the Solar System orbit their primary planets at various distances, and the trajectories of these orbits are differently shaped. To help the readers to better visualize the position of these satellites, I sketched Fig. 2 to exhibit the orbit of satellites according to their planets.

Because I already amply described the subatomic particles in my book on the origin of the chemical particles, and because I tried to make this book as short as possible instead of loading it with many scientific details, I decided not to detail them here. I also studied minerals, rocks, and lava. Particles and celestial bodies in the universe are characterized by different variables (some of which I will explain later in this book): axial tilt, density, eccentricity, escape velocity, gravity, orbital inclination, kinetic energy, mass, motion, orbital period, orbital speed, radius, rotation period, rotational speed, semi-major axis, volume, etc. In *"Turbulent Origin of the Universe,"* I studied hundreds of other variables that, due to space constraints, I cannot address here. Among the variables used to characterize atoms, molecules, minerals, and rocks (details can be found in my book on the origin of chemical particles), I chose to focus on atomic number, abundance, mass, radius, name, and state. For each of these variables, I conducted some statistical analyses that, for the sake of space and relevance, I will not present in this book. For the sake of time, I did not review all of that data here, but it can be found in *"Turbulent Origin of the Universe".* I felt I'd better introduce them here, so those who are curious to learn more know where to find them. Otherwise, I would have had to deal with more than 1,000 pages for this version of the book, which I think might bore readers who may be interested only in the biblical account of the formation of the universe rather than the details some scientists may seek. When studying interactions between particles and celestial bodies in the universe, scientists often use the fundamental forces of nature. In this book, I will focus on gravity, the best-known of the so-called fundamental forces that scientists think govern all interactions in nature.

As I studied the aforementioned variables, I noticed similarities and differences in how they related to one another. When I carefully investigated how those similarities and differences varied across certain ways of clustering of the bodies, many laws appeared to me as the foundation of the structures and behavior of things in nature. By the time I pulled together those trends, I came up with a picture of the origin, not only based on the scientific data but also based on details provided in the word of God, which is filled with facts in plain sight but whose understanding or level of depth has been sabotaged by the mindset, limits, and spectrum of knowledge and discernment of beings that have been influencing human beings. I also detected that in each system, some bodies (primary bodies) are usually positioned at the "center," while others (secondary bodies) orbit or turn around them. Secondary bodies that are closer to their primary body generally orbit faster than those that are farther away. I noted that the primary bodies in each system are usually the largest, while the largest secondary bodies are not always the closest to the primary body. The largest secondary bodies are neither the closest nor the farthest; they are sometimes located between the primary bodies, closer to the primaries than to the outermost bodies. I noticed that most of the system's energy is usually found in the primary bodies. Likewise, most of the energy of the secondary bodies is usually found in the region encompassing the largest secondary bodies. I observed that the shape of the systems of bodies and even of the single bodies has many things in common, and the state of matter in these bodies points to how the ways that some bodies were compressed or wrapped affected the destiny of the bodies born from them. I came to realize that turbulence was the underlying phenomenon present at the beginning of the universe, and that it can holistically explain the distribution, organization, and structures, as well as most of the characteristics, of bodies in the universe.

Looking at the potential application that a breakthrough in turbulence can have in science, I personally think that God, the Creator of all things, made turbulence to be so complex so that human beings do not easily understand how He created the world, and start trying to replicate certain things, which in the end, will cause more trouble in this world. The more scientists understand turbulence, the more they can reverse-engineer many things, unfortunately, without considering their impacts on others. God allowed some of those mysteries to be revealed to some believers, including me, Nathanael-Israel, because He trusts me to make good use of them, including publishing what can be published at the right time. For if scientists properly understand how matter was formed and how they can manipulate it, God may have to end this world quickly so that human beings do not destroy it with their photocopy of the creative genesis of all things. Otherwise, God may have to turn off certain properties of the natural laws governing the universe so that what scientists may have known and tried to apply to destroy the universe no longer holds. For all it will take to change physical laws is just to change some parameters, suggesting that no matter what science may know about creation, God is (and will still be) in control of the universe. This gives me confidence that even when what God has been revealing to me about the origin of the universe falls into the wrong hands of evil-intended people, God is still able to change things in the right direction. For He is the master of all.

If God has made the eyes of every human being to be as accurate as those of some prophets, those eyes could have made human beings able to see clearer than the most sophisticated microscope. But God did not want that design, for He knew that if human beings could easily understand certain things, they could have tried to easily destroy His plan and design to some extent. It took more than 400 years, from the days of Da Vinci (when turbulence studies began), before scientists could begin to see coherent structures in turbulent flows. For 400+ years, they thought turbulence was just disorder and chaos.

Not knowing that God is the expert in turbulence, some scientists even dared to say that God knows nothing about it, just as one of them naively exclaimed: *"When I meet God, I am going to ask him two questions: why relativity and why turbulence? I really believe he will have an answer for the first"* (Werner Heisenberg). This statement of the scientist may be caused by the fact that turbulence modeling is one of the most difficult tasks in modern science. In this book, I will also discuss how turbulence applies to the formation of the universe.

Early in my investigation in 2013, I realized that it would be impossible to explain the origin of the universe by just sticking with linear reasoning in separate fields of knowledge, but by strategically considering facts in various fields of knowledge, such as biology, chemistry, mathematics, philosophy, physics, and the spiritual or invisible world. Although I addressed all of these disciplines in my book titled *"Turbulent Origin of the Universe"*, here, I emphasized the biblical aspects more. For this current book seeks to:

- honor God for His creative work
- help the believers to understand the creation story recounted in the Bible so that they keep believing in God
- help the unbelievers to understand that science does not contradict the Bible if scientific data are properly interpreted, and lead them to the God of creation who can save them.

Throughout my writing, wherever you see "universe-origin", please know that I meant "origin of the universe" or "the origin of the universe". Likewise, wherever you see "life-origin", please understand that I meant "origin of life" or "the origin of life". In the same manner, wherever I mention "chemicals-origin", please know that I am referring to "origin of chemicals" or "the origin of chemicals".

'Science180 Academy' Success Strategy:
SCIENCE180 ACADEMY OVERVIEW

Science180 Academy is a training, speaking, consulting, and mentoring program designed to groom and empower people of all backgrounds in the truth about the origin of the universe, life, and chemicals. According to their background and interests, trainees are taught at different levels of scientific concepts to gain a deeper understanding of the origin of the universe and how to think critically to unearth mysteries hidden in the massive scientific data collected across the globe, which is unfortunately under analyzed. If you want to be enlightened and equipped so you can cause positive changes in your respective field of expertise, then the Science180 Academy program is for you.

Science180 Academy does not confer college credit, grant degrees, or grade its attendants, participants, or students. It is not an accredited university or college, but is the one-stop destination for universe-origin, life-origin, and chemicals-origin experts. It is where scientists and laypeople get all their origin-related questions properly answered. It is the only place where the accurate interpretation of the universe-origin, life-origin, and chemicals-origin data matters a lot.

Science180 Academy brings together Dr. Nathanael-Israel Israel (the Founder of Science180) and other experts to deliver outstanding value, insights, and lessons to help you accurately understand the true origins of the universe, chemistry, and life, so you can tap into that knowledge to improve lives forever. Nathanael-Israel's goal is to give you practicable and undeniable proofs of the formation of the universe so you can be fired up to become the best version of you, and to cause positive changes to your initiatives that will profit you today and forever. For Nathanael-Israel, decoding the origin of the universe and everything in it is not a job, but his life mission, and helping others to fully understand that is his mission. Visit Science180Academy.com today to start.

Science180 improves their knowledge, experience, performance, and answers their questions (related to the universe's origin, life's origin, and chemicals-origin) by crafting a personalized program that perfectly matches their interests, needs, and things that are dear and meaningful to them, whether it is to:

- Become the leader that captures the heart of your followers, prospects, and customers, craving for an unconventional explanation of the origin of the universe, life, and chemicals
- Benefit from continual updates and assistance during your journey to decode the universe, and clear your way for the universe-origin-related freedom, power, technology, innovation, and breakthroughs of the future.

- Bypass technical knowledge that restricts non-experts from accessing the origin-related truth contained in the massive scientific data, and get to the bottom of scientifically locked origin-related secrets regardless of your background.
- Challenge the cosmological status quo and embrace the real change that will disrupt the cages that were holding you
- Connect with practical tips about how to decode the origin of the universe, life, and chemicals and protect yourself from wrong theories in the literature and the media
- Enjoy multiple origin-related programs and choose the ones that best suit your needs
- Fearlessly push the boundaries of human abilities to properly understand what is perceived as un-understandable, mysterious, supernatural, unimaginable, impossible, and unthinkable that holds you back
- Free yourself from boring explanations of the origin of the universe, life, and chemicals and embrace the proven theory that opens doors to unparalleled opportunities
- Get inside secrets about how to locate flaws in origin-related theories so you can save time, money, and other resources to improve lives
- Have reliable access to the world's authority on origin-related matters and get your origin questions professionally answered with the truth step-by-step
- Learn how to daringly contribute to setting on fire all false universe-origin theories and life-origin theories that are enslaving humankind
- Make a difference and blaze new trails for those who depend on your leadership
- Protect yourself and your loved ones by keeping all of you secured and empowered with the true knowledge of the origin of the universe
- Satisfy your burning desire for freedom from beliefs and scientific theories about the universe's origin and life-origin that suffocate you and bind your mind, faith, unbelief, heart, and education
- Scientifically know how God created the universe, and who God is
- Stand as the lightning bolt that electrifies your colleagues who are still struggling to understand the universe's origin.
- Ultimately, boost your confidence in detecting, confronting, and avoiding wrong theories by knowing the facts and processes involved in the formation of the universe

To register or to learn more, visit Science180Academy.com today.

CHAPTER 2

WHY CHRISTIANS DON'T HAVE TO COMPROMISE WITH EVOLUTIONISM TO CONVINCE ANYONE ABOUT HOW GOD DESIGNED CREATION

Throughout this book and to the extent I was authorized, I cracked some of the creation codes hidden in nature. Understanding God's design is also key to understanding how the world was formed and how it has been functioning since then.

2.1. What was God doing before creation?

On May 18, 2016, I was inspired to write this segment about what God was doing before creating the world. For it appeared to me that God did not just get up and start creating things at random, but that He strategically planned everything. Before creation, God was alone in the universe, which originally consisted solely of spiritual things, and He was the only spiritual being. Throughout the time God existed by Himself before creating the physical world, He was alone in the universe, and no spiritual beings had been created yet. By the way, to be precise, I should not have used the pass tense (e.g. existed) to express a past of God, for God is I AM. But I have to use the past tense so some people can comprehend the story chronologically. Before creation, God already had his own comfort and system of maintenance, happiness, provision, rest, etc. As a being who is 100% satisfied and who never changes, God was not surrounded by anything else before creation. Anything that exists in the universe, including the circumference of God or His dwelling place as of today, was designed and created by Him.

Before creation, God used to go about in the invisible spiritual world. The spiritual world contains real things that a mere human being cannot see. Then came a time when God conceived the thought of creating physical things. Therefore, He planned the foundation of creation according to His desire and own counsel. Because He is omniscient and omnipotent, He did not struggle in His planning. As a master planner, He decided to create diverse types of living and nonliving things in an order that would also sustain His will, regardless of what would happen after creation. In other words, God created the universe according to a well-planned process sustained by laws that continue to maintain the things contained in it. Before creating the world, God had planned the mission and organization of everything He was going to create. Being highly organized, God also planned to create a very ordered world.

God did not plan to form everything at the same moment, but after instantaneously creating the initial matter out of nothing, He planned to use a process to shape it into different things. In other words, the process that formed the universe and encodes its current characteristics was not random but intentionally designed by God to serve His purpose. That is why natural laws are not random, and nature does not work anyhow, but according to a set of plans, some of which scientists would later discover and try to use to make artificial things—most of which have been creating trouble for living things that God created according to natural processes, which human artificial thinking does not allow or respect.

God planned to create angels, plants, wild animals, and human beings. He also planned to create an environment that would host and fit each of them. For living things that will stay in the same environment, He designed a way to let them interact with one another through chains of actions and reactions that ecologists have helped unearth. In other words, God's overall plan considered the hierarchy, harmony, interconnectedness, and dependence of everything He would create. His initial plan was to create a universe that would contain Earth in a location compatible with human beings and other living and nonliving things that would also live on that Earth. Because God knows Himself very well and the design He wanted to give to human beings so they could be made in His image and likeness, He therefore designed a process to create a world using a system that would allow Him to have a habitable, comfortable Earth for human beings. In other words, because His goal was to make man in His image, He defined and made things to meet their needs. He designed the chains and all the factors that would be involved in the maintenance of human beings and the other things He was about to create. He also decided to create spiritual beings to minister to human beings and also to be in His presence. God Himself does not need angels before being who He is, for He was also satisfied by Himself throughout the ages that preceded the creation of the universe.

Another key part of God's design is that He wants to be the supreme being who controls everything. Although He gave human beings free will, He can change their will if He wants. For this age, He is letting people and spiritual beings do what they want (to some extent) until the judgment day. In other words, although God wants to control everything, He willingly chose to temporarily let certain things play out without forcing His will on human beings and certain spiritual beings. Those who do not understand this temporary plan think that God does not exist or blame Him particularly for all the bad things happening in the world, which are not always caused by God but by people using their creative minds, which are usually making things not perfectly match God's will. Nevertheless, God planned to control natural laws and the extent to which human beings can affect them. For instance, He causes and controls precipitations (Job 38:25-30). He controls the zodiac and the laws of the sky (Job 38:31-32). He can even speak to the clouds and lightning (Job 38:35). It is not by chance that God placed Himself above everything, including above all of the heavens!

Once God finished and approved His plan, He then implemented it by creating the things He wished. Here, I will review some of the things I believe God planned before beginning creation. Understanding God's plan can deepen the reader's understanding of the creation, formation, interactions, and functioning of living and nonliving things in the universe.

2.2. Why did God create the universe?

Although human knowledge is limited and cannot comprehend all the details of God's motivation for creating the world, I tried to address a few reasons here. Just as a human being, God also has desires, and He does what His soul desires (Job 23:13). In contrast to men, God's desires are always pure and holy. He created the world for his own pleasure and according to the counsel of his heart. He created everything by Christ and for Christ (Colossians 1:15-18), meaning that giving every created thing to Jesus was part of the reason God created the universe. Each creature has a purpose. For instance, He created the Earth not to be chaotic or desolate but to be inhabited: *"For thus saith the Lord that created the heavens; God himself that formed the earth and made it; he hath established it, he created it not in vain, he formed it to be inhabited: I am the Lord; and there is none else"* (Isaiah 45:18). That is why the environmental conditions on Earth match the requirements for life. God did not create the universe by chance but has a plan that He has been fulfilling (Isaiah 46:10-12). It is important to know that as God was planning to create the universe, He also envisioned its end. In other words, after the creation purpose is fulfilled, God will end the current matter. The current chaotic state of certain things in nature is not a failure of God and is nothing that surprised Him, but a direct component of His plan and will to work with His creation until the appointed end comes.

God also created the world so that human beings can enjoy it. That is why the landscape is beautiful, and human beings can appreciate it by traveling to different places or by observing the changes in nature with the seasons and over time. On other planets, things are not as diversified as on Earth. As of now, human beings are not known to inhabit any other planet or celestial body. Likewise, Earth has conditions that meet human needs. God had to create the earthly conditions to be so; else, human beings would have to strive to live in an environment that could not support them. For this alone, human beings are indebted to God for His provision. On Earth, we breathe air freely, but if human beings were put on other planets, they could not survive and would need to import air. This is also true for many other things available on Earth and indispensable for human existence. Even on Earth, some people had to get sick and be unable to breathe on their own before they realized the free gift of a nose and atmospheric air. I could also illustrate that the position of the Earth from the Sun is such that human beings can live and enjoy the Earth, else they could have already died from the strength of the Solar radiation, and the cold had the Earth been placed a little farther in either direction.

Seeing the problems in the world, I wondered on December 31, 2013, why God planned a world that He knew would disobey Him and be in trouble? For most things on Earth seem to be going toward a worse state: ecosystems are being devastated, floods, earthquakes, and wars are increasing, hunger is growing, pollution is growing in cities, people are more inclined to sin than obey God; natural things are being destroyed and replaced by artificial things that are threatening human health, etc. The same day I asked that audacious question, it appeared to me that, as He was planning, God did not intend everything to stay perfect forever. He knew certain things would go wrong as far as human beings and some angels would be concerned. In other words, God did not plan to have human beings and all angels act like robots that would be like puppets or marionettes in his hands and never fail or disobey Him. Another way of saying this is that God did not plan to have all of His creatures be like robots, which will mechanically execute behavioral commands like some celestial bodies, which have been orbiting their primary bodies for thousands of years without ceasing.

It appeared to me that the matter is obeying natural laws set by God, and most of the natural laws connected to nonliving matter have not changed beyond the devastating impact the human beings and fallen angels have been inflicting on it. At least, it seemed to be obeying God more than human beings, for few things have changed in them than with human beings. In other words, natural laws connected to nonliving things have not changed as much as those connected to human beings and the interactions between them, therefore suggesting that the biggest problem in and of the universe is with human beings and the evil spiritual agents that negatively influence them, but which are invisible to mere human beings. Because all of these problems could not have been unknown to God beforehand, I felt that some of the things God created were meant to go through a selection process, after which a separation would occur, and retribution would be given accordingly. Human beings have been passing a test of life whose outcome is well known to God, even before He created everything. That is why God designed the plan of redemption of humankind even before He laid the foundation of the world, and crucified Jesus before the world nailed him to the cross.

Therefore, as much as it is important to understand how God created the world, it is also imperative to know that God knew what His creatures would do even before He created them. Therefore, He also took into consideration all the actions and reactions of everything He was planning to calibrate His universal plan before He launched it. Hence, He is the Alpha and Omega, knowing the end before the beginning. In contrast, all other things in the universe do not know properly how they began and how they will end. Likewise, the way God has been responding to things and beings in the universe has been taking into consideration not just their instantaneous needs or desires alone, but ultimately His global plan, which goes beyond space and time and the imagination that anything in the universe could have. That is why those who reject God because they focus on their needs—as if God were there to solve all the problems of humankind—are just unaware of how God operates. Likewise, even when the believers ask things that are not aligned with God's universal plan, He does not answer. In other words, the way God responds to the needs, desires, and aspirations of His creatures is well aligned with the reasons He created the universe and the whole, big picture of His plan. Otherwise, because the created things and beings do not fully know God nor comprehend His language, the satisfaction of all their requests could not be coordinated with the divine plan, which is everlasting, pure, complete, and unchangeable!

Therefore, instead of trying to fully understand or argue with God and His design, human beings are better off seeking His ways. Job tried to argue with God, but when God showed him the power of creation (Job 38-39) and the motivation behind certain things He created, Job had to shut his mouth and confess his ignorance (Job 40: 1-5). Instead of arguing about things in nature and God, human beings would be better off worshiping Him just as Apostle John did in Revelation 4:11: "*Thou art worthy, O Lord, to receive glory and honor and power: for thou hast created all things, and for thy pleasure they are and were created*".

2.3. Measurement of the foundation of everything before creation

As I was studying the Bible's Book of Job on December 18, 2013, meaning about 7 years before I wrote this chapter, I learned some details of how God designed everything. God had defined the dimensions of the Earth and laid its foundation (Job 38: 4). Then He stretched the measuring lines between the Earth and the planets, satellites, and stars (Job 38:5) so they could be placed in their proper places. At one point, God shut up the sea behind doors (Job 38:8) so it would not violate its boundaries. I later realized that if the size of the Earth were larger than its current size, the length of the day and night would have been different. Consequently, human behavior could have been different. The measurement that God did before creating the Earth was aligned with His plan, which cannot fail. Prophet Isaiah went further to explain that God measured things with His hands: "*Who hath measured the waters in the hollow of his hand, and meted out heaven with the span, and comprehended the dust of the earth in a measure, and weighed the mountains in scales, and the hills in a balance?*"

(Isaiah 40:12).

This verse suggests that the measurement that God did was not just about length, but also about weight, density, particle composition of the heavens (not just the Earth), size of the mountains and hills, etc. As I explained later in this book, the position, speed, movement, and many other characteristics of celestial bodies and even of chemical particles (e.g., atoms and subatomic particles) follow specific laws highly connected to their distance from one another or from their primary bodies they orbit. All of these measurements were part of God's plan before creation, so they matched the requirements of men and other living things and even nonliving things on Earth and elsewhere. In the chapter reserved for semi-major axis, mass, energy, and speed, I will provide specific details about these variables, illustrating God's measurement and the laying of foundations.

2.4. Design of the foundation of the heavens and earth

Early in the beginning of the creation of the universe, God laid the foundations of everything, including the foundation of the heavens and the earth:

- "*And, Thou, Lord, in the beginning hast laid the foundation of the earth; and the heavens are the works of thine hands*" (Hebrews 1:10).
- "*Of old hast thou [God] laid the foundation of the earth: and the heavens are the work of thy hands*" (Psalm 102:25).
- "*Who laid the foundations of the earth, that it should not be removed forever*" (Psalm 104:5).

Prophet Isaiah went further to explain that God made the foundations of the earth using His own hands: "*Mine hand also hath laid the foundation of the earth, and my right hand hath spanned the heavens: when I [God] call unto them, they stand up together*" (Isaiah 48:13). Much was not said about what that foundation consisted of. However, when the design of a building's foundation of a building is considered, builders consider many things, including the following:

- Design of the building
- Geographical location (God is on top of the universe)
- Climate
- Soil
- Moisture
- Budget (which affects the quality, materials used, and the timing of the construction)
- Nature of the materials to be used for the building
- Accessibility of the building

For the foundations of the earth and the universe, God would have addressed more than those points mentioned above. Considering the processes of the formation of the universe, the foundation laid by God could have involved not only the nature of the "turbulent prima materia"—the initial material or substance that God created out of nothing before using it to form everything else—, but also the measurements and all other details of the design God planned before starting to build the universe. Throughout the psalms, King David, the man after God's own heart, also acknowledged on many occasions the great care God took in creating the universe and how it declares His glory. The marvelous things and beings in the universe could not have been formed without the solid foundation that God laid. God used His wisdom and knowledge to divide the precursors of the universe to establish the foundation of the earth and to form different types of precipitation (Prov 3: 19-20).

2.5. Encryption of the creation code in natural laws

As part of God's plan and goal, human beings (and particularly the unbelievers) are not supposed to fully understand all of the details of creation. Therefore, He hid many secrets from the beginning, but through His prophets He revealed a brief summary to those who believe in Him. Some of those revelations are found in the Bible and in some lost or rejected Jewish scripts that I chose not to mention here, but *"Origin of the Spiritual World"*.

Considering what the scriptures revealed about creation, I realized that God did not really hide the story of creation, but it is human beings who have been refusing to understand the Genesis story from the true perspective, instead of from that which they think fits their thinking or imagination. If certain scriptures had not been rejected by those who compiled the Bible and later removed from the most popular translations, people would have been more aware of many specifics God has already shared about how He created everything. Later in this book, I explain why it was better that the code of the creation of the universe was encrypted into the requirements God placed on the blessing He promised to His people. In other words, it must be impossible for human beings to understand how God created the universe if they do not want to wholeheartedly honor and serve Him with the things they will do once they decrypt the universe's code. That is another reason I think the discovery of the code governing the formation of the universe is a sign that we are nearing the end of a major era and the beginning of another I presented in my book on the age of the universe.

I also came to realize that if all proofs of creation had been revealed and detailed to human beings 100%, faith, which is the barometer of selection or approval for those who believe in God, would have lost its value or meaning. For if everything could be known or demonstrated physically, there could not be any room or sense to believe. In other words, the lack of a demonstration of everything God has done

23

Science180: Accurately Understand Universe-Creation. Be Happy Forever!

is a reason for faith to endure throughout the ages, until everything previously hidden is revealed. That is why, although many prophets have existed since the beginning of the world, God did not spend most of His time trying to use them to demonstrate to the world all the mathematics or physics He used to create and maintain it. Otherwise, it could have been as a teacher answering all the questions that he would ask on an exam, and after ensuring that the students understood the questions and the correct answers, the teacher went ahead to let the students take a test based on the exact questions. In such a case, the teacher could miss the point, and the examination could be meaningless. Likewise, before the end of time or the due time, God cannot show human beings all the evidence of His creative works. For life is meant for people to choose to live or die as they wish, so they can live forever with Him in heaven or away from Him in hell according to their earthly choices. Hence, from the beginning, God prohibited Adam and Eve from seeking knowledge, for the acquisition and use of knowledge cannot solve human problems in agreement with the natural laws and plans that God has for the universe and its content. Hence, those to whom God can confide His secrets are not always allowed to reveal everything. Likewise, I do not feel comfortable revealing to the world everything that I know about how God created the universe, for there are things not meant to be blown to the whole world at this moment yet! Toward the end of this book, I have a separate chapter on why God hid secret codes of creation from human beings and even from (some) angels. I reserved some details for later because their understanding requires prerequisites I must cover in the next chapters before returning to them.

As I close this segment, please remember to use faith to demonstrate what you do not understand: *"Now faith is the substance of things hoped for, the evidence of realities not seen. For by it, the elders received commendation. By faith, we understand that the universe was created by the word of God, so that what is seen did not come from anything visible"* (Hebrews 11:1-3, Tree of Life Version). Once God finished planning and designing, He then implemented His plan by beginning to create.

After God designed everything according to His own will, without consulting anyone or anything to counsel Him, He implemented His design. The following chapters will demonstrate how the universe could have formed.

Another Book by Nathanael-Israel Israel:

FROM SCIENCE TO BIBLE'S CONCLUSIONS

THE # 1 UNIVERSE-ORIGIN MASTERPIECE OF ALL TIME … AND THE MOST ACCURATE SCIENTIFIC FORMULA THAT STOOD AND WILL STAND THE TEST OF TIME AND OF MATHEMATICS

The real reason scientists have been struggling to accurately understand the universe formation is that they have spent centuries collecting expensive, complicated, and massive amounts of data, but learned very little, if anything, about how to step back unconventionally to properly analyze it to decode the universe. Consequently, people learned to collect all kinds of data everywhere to build models and imaginary concepts that betray their discernment, but they never learned to unlearn wrong theories, nor learned how to stop trashing great raw data hidden in theories they dislike or misunderstand; they never knew where to find and how to properly combine the fundamental variables without which it is impossible to ever clear the way so their data can properly work for and precisely lead them to the real origin of the universe. How can people abandon the dangerous theories they think are correct because they don't know any better ones?

Lucky you, that is where Dr. Nathanael-Israel Israel, the founder of Science180 (Science180.com), came in to properly reanalyze and put these costly, underrated data under control to provide the accurate, simple solution people have been looking for throughout the ages, but that they have ignored.

In *"From Science to Bible's Conclusions,"* you will:

- Get a world-class explanation of the 4 fundamental variables without which it is unquestionably impossible to ever decode the universe-formation scientifically
- Save time and money, and enjoy a life filled with the wonderful peace that the accurate understanding of the universe's origin can create
- Discover the errors in the scientific theories and religious belief systems about the universe-formation that are putting you at risk, and learn how to take control over cosmological threats lurking at the edge of your rational mind, faith, disbelief, or doubt
- Unlock the accurate scientific formula to rationally test the existence of God in a historic way that uncompromisingly satisfies both believers and skeptics (Science180.com/public)

- Get all you need to become a knowledgeable person who will never again need anybody else to explain to you the origin of the universe, for you will fully understand and articulate it yourself, and rationally know whether science is really at war with religion
- Receive deep insights that even those who went to university for years were not able to decrypt by themselves, so you can equip yourself to eliminate all forms of scientific and religious universe-origin prejudices
- Discover whether the scientific data finally confirms that the formation of the Earth was completed on the 3rd day, while that of the Moon and the Sun was on the 4th day of creation, like the Bible says, or whether the data proves that it took billions of years to progressively form the universe
- Understand the celebrated scientific formula that rationally puts to rest all debates about the relationship between science, faith, and all theories about the universe's origin, so you can properly develop yourself, expand your network, and shape your future

Quickly grab and read this scientifically verifiable, bestselling book to finally get the accurate, jaw-dropping answer that has been rationally shaking believers, skeptics, and all freethinkers. Don't wait!

Dr. Nathanael-Israel Israel has had the honor to be acknowledged as the #1 universe-origin, life-origin, and chemicals-origin expert.

He is the author of "Turbulent Origin of the Universe," "Reconciling Science and Creation Accurately", "Turbulent Origin of Chemical Particles," "Turbulent Origin of Life, "How Baby Universe Was Born", "Science180 Accurate Scientific Proof of God".

Visit Israel120.com to learn more about this world's most trusted expert who helps scientists and laypeople to properly decode the origin and formation of the universe, life, and chemicals, so people can live more effectively nonstop.

CHAPTER 3

WHY DO PEOPLE NOT HAVE TO DENY GOD AND EMBRACE SECULAR THEORIES IN ORDER TO EXPLAIN THE UNIVERSE'S BEGINNING, BECAUSE THEY THINK THAT IT IS IMPOSSIBLE FOR SCIENCE AND FAITH TO MEET?

More and more people are denying God to embrace secular theories because they think that it is impossible for science and the Bible to rationally explain anything about the universe's origin. I am here to tell the whole world that this trend is tremendously dangerous and how we can start fixing it, while explaining the beginning of the universe.

3.1. Codes of the formation of the universe encrypted in the Judeo-Christian creation narrative

When people talk about the beginning, they usually seem to forget most of the things I addressed in the previous chapters, for they are not plainly understandable by a mere human being, who did not do their homework in going deep into the search for the truth! To confidently support the inspiration that I had and the discoveries I made during my "investigation" of the origin of the universe and during the writing of the corresponding manuscripts, I had to rely on some key irrefutable books accepted by Christian, Jewish, and Messianic communities.

Although many biblical books contain information about creation, when the beginning of the world is mentioned, two references are frequently used: Genesis 1 (written by Moses) and John 1 (written by Apostle John). Other biblical books that contain key information about the creation of the world include Job, Psalms, Isaiah,

Jeremiah, and Revelation. In the chapters to come, I will often refer to many scriptures from these books that support the creation of the universe. Beforehand, I decided to list some of those references here so that, when I cite them later without always writing them out in full, readers can easily refer to them if needed. Presenting these references here can also help readers understand and compare the context in which those verses were originally mentioned with the context of the story I crafted in this book. Although I did not detail or comment on all of the verses in Genesis 1 in this chapter, I decided to quote those related to what happened during the first 4 days of creation, for I will use them later in other chapters as needed:

> Genesis 1:1 *In the beginning God created the heaven and the earth. 2 And the earth was without form, and void; and darkness was upon the face of the deep. And the Spirit of God moved upon the face of the waters. 3 And God said, Let there be light: and there was light. 4 And God saw the light, that it was good: and God divided the light from the darkness. 5 And God called the light Day, and the darkness he called Night. And the evening and the morning were the first day. 6 And God said, Let there be a firmament in the midst of the waters, and let it divide the waters from the waters. 7 And God made the firmament and divided the waters which were under the firmament from the waters which were above the firmament: and it was so. 8 And God called the firmament Heaven. And the evening and the morning were the second day. 9 And God said, Let the waters under the heaven be gathered together unto one place, and let the dry land appear: and it was so. 10 And God called the dry land Earth; and the gathering together of the waters called he Seas: and God saw that it was good. 11 And God said, Let the earth bring forth grass, the herb yielding seed, and the fruit tree yielding fruit after his kind, whose seed is in itself, upon the earth: and it was so. 12 And the earth brought forth grass, and herb yielding seed after his kind, and the tree yielding fruit, whose seed was in itself, after his kind: and God saw that it was good. 13 And the evening and the morning were the third day. 14 And God said, Let there be lights in the firmament of the heaven to divide the day from the night; and let them be for signs, and for seasons, and for days, and years: 15 And let them be for lights in the firmament of the heaven to give light upon the earth: and it was so. 16 And God made two great lights; the greater light to rule the day, and the lesser light to rule the night: he made the stars also. 17 And God set them in the firmament of the heaven to give light upon the earth, 18 And to rule over the day and over the night, and to divide the light from the darkness: and God saw that it was good. 19 And the evening and the morning were the fourth day.*

For now, I stopped on Genesis 1:19 because, to explain the formation of the celestial bodies, I did not need the other verses yet. At his turn, Apostle John, known as the best friend of the Lord Jesus, started his narration of the creation (that he had likely heard from Jesus) like this:

> John 1:1 *In the beginning was the Word, and the Word was with God, and the Word was God. 2 The same was in the beginning with God. 3 All things were made by him; and without him was not any thing made that was made. 4 In him was life; and the life was the light of men. 5 And the light shineth in darkness; and the darkness comprehended it not. 6 There was a man sent from God, whose name was John. 7 The*

same came for a witness, to bear witness of the Light, that all men through him might believe. 8 He was not that Light, but was sent to bear witness of that Light. 9 That was the true Light, which lighteth every man that cometh into the world. 10 He was in the world, and the world was made by him, and the world knew him not. 11 He came unto his own, and his own received him not. 12 But as many as received him, to them gave the power to become the sons of God, even to them that believe on his name: 13 Which were born, not of blood, nor of the will of the flesh, nor of the will of man, but of God. 14 And the Word was made flesh, and dwelt among us, (and we beheld his glory, the glory as of the only begotten of the Father,) full of grace and truth. 15 John bare witness of him, and cried, saying, This was he of whom I spake, He that cometh after me is preferred before me: for he was before me. 16 And of his fulness have all we received, and grace for grace. 17 For the law was given by Moses, but grace and truth came by Jesus Christ. 18 No man hath seen God at any time, the only begotten Son, which is in the bosom of the Father, he hath declared him.

Before I started decoding or breaking down the scientific implications of the encrypted creation narrative contained in the above-mentioned verses, I would like to mention that other books known for thousands of years also contain key information about the origin of the universe. For instance, during my reading and investigation of the Pseudepigrapha (the lost and rejected scriptures), which was a key component of my morning readings in 2020, I also came to realize that many of them contain authentic stories about creation that do not contradict the biblical account but instead bring some light onto things not detailed in the Bible. Some of these books are the Books of Enoch, the Book of Esdras, the Book of Jubilees, the Book of Jasher, the Books of Adam and Eve, and the Gospel according to Mary Magdalene. Among these books, the 3 versions of the Books of Enoch are the most revealing and shocking. Because of the depth of the information in some of those books and the fact that they align with biblical stories and scientific data, I felt obliged to briefly mention them here before detailing some of them in my book, *"Origin of the Spiritual World."* For instance, although less known and quoted by Christians and Jews, the Books of Enoch are filled with deep details about the formation of the universe and the beings it contains, including angels, etc. Nevertheless, I refrained from addressing those data here, and those who are interested can refer to *"Origin of the Spiritual World,"* the pseudepigraphic version of this book.

Because most of the stories revealed in the Bible and in the Pseudepigrapha were written by prophets to whom God revealed things, and considering the prophetic movement, which is being raised up in these last days, and how current revelations through contemporary prophets are ignored by some people, I felt compelled to search and investigate what some renowned contemporary prophets are saying about the creation of the universe. Toward that end, in the last trimester of 2020, I visited the YouTube accounts of some of these prophets. Although I respect all of these prophets and learned many things from them, I was shocked to realize that some of them are unable to properly interpret the creation story of Genesis 1, and instead landed their teaching in evolutionism and Old-Earth creationism, a creationism according to which the earth, the universe, and their content or inhabitants were created millions (or billions) of years ago instead of in six 24-hour days! I am not saying this to dirty any prophet or any other man of God, but to show how prophesying is different from interpreting a prophecy, and also to show that no human being knows everything. When I explored the narrative of those who believe in Young Earth creationism (creation in six 24-hour days), I also found many errors that need to be addressed to align the creation message with an unquestionable reality. In this book, I did not point fingers at anyone for their false interpretation of the creation narrative in the Bible's Book of Genesis, but I just used the revealed knowledge and the scientific data to clarify the depth of what God has revealed and what science has discovered (although most scientists are unable to properly interpret their data). Below, I expose some of the 3 fundamental mistakes people have been making in properly interpreting the creative narrative in Genesis and in other biblical books.

3.2. Fundamental mistakes preventing people from decoding the Biblical creation story

Here, I will focus on three fundamental mistakes that have been preventing people from understanding the creation story of Genesis 1: the precursors of the bodies and the correct chronological timing of the events. Indeed, when I reviewed the stories about the creation of the universe, even those accepted by many people who claim to believe in the Judeo-Christian account of creation, I realized that some of the most erroneous ones make 3 mistakes, such as the following:

- Failure to distinguish between the precursor of a body and the "finished" or "complete" version of that body mistakenly caused some people to think, for instance, that the first verse in Genesis is talking about the "complete" or "finished" version of the world that God created;

- Distrust in the chronological account of the story in Genesis 1, making people reorganize the story of Genesis 1 as they please, wrongly thinking that Moses was not smart enough to state things according to the order they were created or formed, or incorrectly thinking that God did not recount a chronological story to Moses;

- Distrust in the timing that was revealed to Moses and that he wrote down with a detailed account of the happenings in each of the 6 days of creation, which some people unfortunately rejected and found themselves with theories attempting to explain the creation of the universe as a process that took millions of years instead of the 6 literal days emphasized in Genesis 1.

Indeed, many verses on the creation of the universe as recounted in Genesis cannot be properly understood if the reader takes them chronologically without keeping in mind that some bodies addressed with the same words were precursors of those bodies whose formation was later finished in other verses. For instance, Genesis 1:1 is a quick summary of the events of the first three days of creation (Genesis 1:2-13). Although certain bodies could have been fully formed by the end of the first day of creation, many others could have needed to go through additional changes before getting their final shapes and characteristics. In other words, the content of the "heaven" mentioned in Genesis 1:1 includes the precursors of bodies, including those whose formation was completed by the 4th day, the date that the formation of the Moon and Sun was finished. Likewise, the "earth" that is mentioned in Genesis 1:1 was the precursor of the earth that was fully formed around Genesis 1:9-13. Similarly, Genesis 2:4-25 provides a detailed account of the creation of Adam and Eve, summarized in Genesis 1:26-27. Unfortunately, some people think that the account of Adam and Eve's formation in Genesis 2 is a story different from that addressed in Genesis 1:26-27. This misunderstanding alone led some people to believe that the Bible contains 2 conflicting accounts of creation. In the chapter on the precursors of the bodies in the universe, I explained how they were shaped into the bodies we have today and how biblical verses also support that process.

Likewise, because people distrust the chronology of Genesis 1, they forge diverse theories to justify their mistakes. By rejecting the chronology of Genesis 1, some people are left with no record to check their own chronology. Therefore, they found themselves with various stories patched just to fit the fictive story that they think fits what they want to say, according to what they think could have happened, but that they unfortunately narrated as things that must have happened. In other words, the rejection of the chronological story of Genesis 1 opened the door to many false theories, even among Christians, that cannot be reconciled with the Bible. Therefore, even those who call themselves believers in God embrace false theories at the expense of the Genesis 1 story, which some people see as archaic, insufficient, pre-scientific, or non-scientific, even though it is the highest level of science. That is why even among the creationists (who fervently criticize the unbelievers for not believing in God), many embrace theories that at their core reject God. For instance, some creationists claim to believe in God, yet they accept some stories that claim that the universe was formed by evolutionary processes that they think took millions of years, an idea contrary to what the Bible says.

As I was working on the scientific explanation of the origin of the universe (see my books *"From Science to Bible's Conclusions"* and *"Turbulent Origin of the Universe"*), I was shocked to discover that even scientific data (e.g., see the chapter on timescale) support that the Moon and the Sun, for instance, were formed on the 4th day as revealed in the Bible thousands of years ago. Given that the King James Version of the Bible dates the writing of the book of Genesis to 1450-1410 BC, it is fair to say that for at least 3500 years God has been telling people how He created the universe, yet most people refuse to believe in Him. If I can dare to talk here about the detailed account of creation revealed in the book of Enoch, people would be shocked about how God has done His best to illuminate human beings with the truth about the origin of the universe, yet some still refuse to believe it.

Finally, because some people think that the days mentioned in Genesis 1 are not literal 24-hour days, they open the door to wrong imaginings that use limited physical means to lead them to believe in long processes of millions of years, thereby distancing them from the truth, which should enlighten them. What did they consider a day then? Would they want to work on one of those long, extended days? Most people don't even want to work 8-12 hour days. I have never seen a day extended beyond 24 hours, and most workplaces would not want to pay their employees for those extended hours. Even if they were to pay their employees for working more than 24 hours on one of those extended days, the Internal Revenue Service (IRS) or the government tax collection agency would take most of it in taxes. We'd better stick to 24 hours in one day. How can someone attempt to demonstrate the truth in Genesis 1 if he or she really believes that the story is random, longer than a 6-day event, or that everything was already formed on the first day? Throughout this book, without distorting the chronological story in Genesis 1, I explained how everything created and formed fit into six 24-hour days.

3.3. Eternity and the eternal realms

As I talk about "the beginning" of creation, it is important to mention that this also refers to the beginning of time known today. For before creation, the world in which God lived and is still living is not and will never be defined in our time. As of today, human beings understand time by referring to the movements of the celestial bodies, particularly the Earth, the Moon, and the Sun. Because none of the celestial bodies existed before creation, the time that existed before creation could not be defined using our terms. In other words, before creation, God had been living in eternity, and even after creation, He is still living in an eternal world that does not change, and where time could not be defined using any change of position of the celestial bodies. This does not mean that God and those who live and will live in eternity will not have a way to know when to do certain things according to a plan. In other words, in the eternal realm, there can be a way to define time using other parameters not comprehensible by human beings.

Therefore, I did not seek to go any deeper into my understanding of eternity. When I one day join God in eternity, I will know, and in the end, the joy and rest I will have will probably cause me to not be seeking any more answers, for beforehand, a lot of hidden things will be revealed. Eternity contains everything in the past, present, and future. Hence, those who live in eternity or can tap into the eternal realms can access information that is unlimited by time, space, or anything else. Hence, nothing else. Those who have a prophetic gift can tap into prophetic information by entering a supernatural realm where eternal information and resources are available on the past, present, and future events or happenings. This

Nathanael-Israel Israel: Acknowledged as the World's Most Trusted Expert That Properly Decoded the Formation of the Universe, Life, and Chemicals implies that the current events in the universe are part of the eternal knowledge that God knew before laying any foundation of the universe. That is also why God was able to declare the end from the beginning (Isaiah 46:9-11).

3.4. Birth or beginning of time

The birth of time, as we know it today, began when the precursors of matter began moving and shaping daughter bodies. Earthly time is usually measured in seconds, minutes, hours, days, and years. As I will explain later, it takes one day for the Earth to complete one rotation around its rotational axis, whereas it takes one year for it to complete a revolution around the Sun. The Moon, which is the closest celestial body to the Earth, takes 27.4 days to finish one rotation around itself and about the same time to finish a revolution around the Earth. The Sun, which is the closest star to the Earth, takes about 25.4 days to finish one rotation around its rotational axis.

According to the Bible and even according to the scientific data (a demonstration is coming up very soon in other chapters), it took more than 3 days for the formation of the Sun and of the Moon to be completed, implying that, unlike the Earth, there could not have been a measuring stick for the first 3 days of creation. In other words, the reference to the rotation of the Earth to measure time is not by chance, but a reality that also matches the chronology of the birth of the celestial bodies. Although many other celestial bodies could have been born earlier than the Earth, none of them is close enough to the Earth to be visible to the naked eye and allow an easy visualization and assessment of time.

While scientists see space as having many dimensions, and most human beings see time as unidirectional (moving from past to future, passing through the present), those who can tap into eternal realms see things unbound by time, meaning they see them from an eternal perspective. And because the amount of eternal information related to anything is huge, no human being can claim to know everything about anything. Hence, even prophecy by the most famous prophets is partial, and those who claim to know everything will sooner or later be shocked at how little they know even about basic things. I believe this is how God keeps us humble in our knowledge. For if people knew everything, they wouldn't be hungry to seek God and other things crucial to sustain life. For even the devil claims to know a lot of things, but he didn't know enough to prevent his own fall from heaven.

Before I close this segment on the birth of time, I must say something about the day, month, and year of the birth of the universe. An outstanding question worth asking is in which month did God create the universe? Although the Gregorian calendar may fool some people into believing that the year begins in January, making some think that the universe could have started in January, evidence passed down to the Jews for thousands of years suggests the opposite. Indeed, since the days of Moses (i.e., more than 3000 years ago), the Jewish culture teaches that the month of Tishri (which is around September) is said to be the creation month of the

Universe. In other words, although the exact date of birth of the universe is not known by human beings, evidence revealed to Jews and encoded in their calendar seems to suggest that the universe was formed at the beginning of Tishri, around September. I was shocked to know that even after the gentiles changed the Jewish calendar and put the beginning of the year to January, the month of Tishri was still about the 9th month, which still alludes to a birth after 9 months. The month of Nissan, which is the beginning of the Jewish religious calendar, is about the 3rd month of the Gregorian calendar (i.e., March), while Tishri (the 1st month of the Jewish civil calendar) is the 6th month in the religious Jewish calendar, a mystery that many people do not understand. As far as the date of birth is concerned, God began creating the universe on Sunday and then rested on Saturday, the 7th day (the original Sabbath ordained by God, it was later that the Gentiles changed the Sabbath to Sunday, yet Jews and messianic believers still observe Saturday as the day of rest). What I said in this paragraph about the month of Tishri and the creation month used to be my ideas until 2024 when I discovered something related to other Jews called the Essenes, which has been making me think that Tishri may not even be the month of creation, for the calendar of the Pharisees may have been corrupted to the point that some references were shifted! Later, I will address the exact age of the universe. Furthermore, I also have a forthcoming book on the age of the universe.

3.5. What and where was the beginning?

The beginning that Apostle John addressed in John 1:1 seems to go further in the past history than the one Moses mentioned in Genesis 1. For Moses started his story without saying anything about the origin of God, while Apostle John went deeper and explained how Jesus (referred to as the Word) came into being, or at least how he became flesh. Who is the creating God that Moses referred to in Genesis 1 and that Apostle John referred to in John 1? The Hebrew word "Elohim," used to refer to God in Genesis 1, is the same as the one used in John 1, implying that although Jesus, as the man, is different from Jesus as God, God as mentioned in Genesis 1 is the same as God mentioned in John 1. Therefore, Jesus is the same God, Elohim, referred to in the Bible. Knowing that Jesus is God, and it is believed that God cannot have an origin that can be explained, it appears that John 1:1 may not be trying to explain the origin of Jesus as God. Given that Apostle John is considered the disciple Jesus liked most, it seems Jesus confided deep secrets to him; hence, John seems to have deeper knowledge of Jesus than the other apostles. In fact, without the writing of Apostle John, questions about the divinity of Jesus could remain unanswered. In other words, Apostle John may have had a deeper understanding of how the world was created since he was familiar with the Torah and readily had Jesus available to ask any questions he may have had.

Knowing that before the beginning of time, there was just eternity, the beginning that Moses and Apostle John talked about in their books is the same and is referred to as "Bereshit" in Hebrew. For eternity cannot be different, and the beginning of time cannot be different. Apostle John and Moses just focused on different things in their narrative about the beginning. That is why I believe that the beginning that they both talk about is the beginning of creation, not the beginning of God.

The scientific and even the biblical evidence I have gathered suggests to me that the current state of the universe was not the state at the beginning of creation. For in the process of creating the world, God shifted or moved the precursor of the universe from a position above to below. In other words, the universe descended from where its precursor was at the very beginning. In addition to the expansion of the universe (which is documented by both scientific and biblical evidence), I cannot rule out that the universe could still be descending to some extent as a system of bodies moving from a higher position to a lower one at the same time that it is expanding. The position where it all started could have been called the beginning. In other words, just as time had a beginning, there was also a region or location in the universe where creation occurred and from where the precursor of everything started the journey that defined the formation and characteristics of everything known in the universe today. Unlike being a single point as some theories falsely claim, the beginning of space and all the matter it contains was a very huge region that my mind cannot even imagine, but that is different from the positions of the things present in the universe today. If I can point to a direction for the position of where the universe started, I could say a position located between the dwelling of God and the current universe containing all of the celestial bodies. In *"Origin of the Spiritual World,"* the pseudepigraphic version of this book (which contains references from the Books of Enoch and others), I provided more evidence on this.

Because God did not want to create a world that would be on the same level as Him, He ensured the world would be very far from Him, and all human beings cannot see Him as they stand on the Earth, but after some time of testing, that will qualify some to dwell in His presence filled with light and fire! Furthermore, as I revealed in *"Origin of the Spiritual World,"* some human characteristics changed after Adam and Eve sinned. For instance, in the Garden of Eden, they could see the angels in heaven, and they were in direct contact with God. In other words, God planned the distance between the Earth and Heaven by considering what would happen when human beings would sin. Hence, in His plan, God intended to have different layers or hierarchies in the world, therefore laying the foundation for what is called "heavens." As I detailed elsewhere, the hierarchy among angels, even among Satan and the fallen angels, and the hierarchy of influence between and among the celestial bodies, evidence the structure or order God planned for the creatures.

CHAPTER 4

HOW TO DETECT AND CORRECT PEOPLE WHO ARE LYING ABOUT THE CREATION OF THE UNIVERSE, WHILE WRONGLY INTERPRETING THE MEANING OF "HEAVENS AND EARTH" IN THE GENESIS STORY

"In the beginning God created the heaven(s) and the earth" (Genesis 1:1). How many times haven't you read that verse? Despite efforts to decrypt its meaning using contemporary scientific terms, many questions remain, and I sometimes feel like God purposely sealed certain creation details until the last days.

- Is the earth mentioned in Gen 1:1 the completely formed earth known as of today?
- What is heaven?
- Are the heavens and earth mentioned in Gen 1:1 the same as the heaven and earth known as of today?
- How many heavens are there in the universe?
- Do the heavens mentioned in Genesis 1:1 include the heaven that is the dwelling place of God?
- Does God live in the heaven(s) mentioned in Genesis 1?
- Is the heaven that Jesus referred to in the Lord's Prayer (Our Father who art in heaven, hallowed be thy name…) the same as the one mentioned in Genesis 1?
- Where was God living before creating the heavens?

Details about the formation of the earth on Days 2 and 3 suggest that the earth mentioned in Gen 1:1 cannot be the complete, well-formed, and well-shaped earth known as of today or by the end of the 3rd day, but a precursor of the Earth.

Although some Bible translations render "heaven" as singular, other versions consider it as plural. Other Biblical verses suggest that God created many heavens, implying that those who consider the "hashomayim" (the Hebrew word used to refer to "the heavens" in Genesis 1:1) as plural cannot be blamed. "Shamayim" means "heavens" while "Hashamayin" means "the heavens;" the prefix "Ha" stands for "the." The Hebrew word "Mayim" means "water(s)." That is why "Shamayin" is considered to be composed of 2 words: the "sham" or "samay" and "mayim", and together they mean "water is there". "Hashamayin" is also translated as "the water (that is) there." Although the etymology of the Hebrew word "Shamayin" can be disputed, it strongly suggests that the precursor of the universe or the precursor of the Earth, or the precursor of the heavens surrounding the Earth, contains water or a water-like fluid. Those who consider the word "hashomayim" as singular may be referring to just the earthly heaven. Before I started diving into the real meaning of heaven(s) and the scientific implications of its mention in the first verse of the creation narrative, I felt like I should first briefly review what the Bible says about heaven(s).

Based on the stories of Apostle Paul concerning his visit to the 3rd heaven, most Christians believe that there are 3 heavens. However, the Jews believe in at least 7 heavens. But according to other Jewish books, there are 10 heavens and the 10th heaven is considered the dwelling place of God. Is it by chance that God gave to the Israelites the 10 commandments and also that the pseudepigrapha suggest the existence of 10 heavens? Other biblical verses suggest that God lives in the heaven of heavens, implying that the things that God created in the universe could fit into 9 heavens. It is not by chance that babies also spend 9 months in their mother's womb before birth. Is it by chance that the month of Tishri (which is around September, the 9th month of the Gregorian calendar) coincides with the creation month of the Universe?

In my search for the meaning of heavens, I came across a commentary that I would like to share. Heaven is also used to describe the sky or atmosphere. For instance, the Orthodox Jewish Bible translated the word "heaven" mentioned in Gen 1:6-8 by the Hebrew word "rakia", which the King James version translated as "firmament," while other versions rendered it as "sky" or "expanse." The word "firmament" mentioned in Gen 1:6-8 is used a lot in the book of Enoch. However, its meaning is not very clear and it is hard to find a contemporary word even in this space age that properly defines it. During my reading of 3 Enoch (the Hebrew Enoch), I found an interesting commentary in 3 Enoch 4:7 about rakia, also rendered as raquia or rakia:

> Raquia is a key Hebrew word in Genesis 1:6-8a. It is translated "firmament" in the King James Version and "expanse" in most Hebrew dictionaries and modern translations. Raqa means to spread out, beat out, or hammer as one would a malleable metal. It can also mean "plate." The Greek Septuagint translated raqia 16 out of 17 times with the Greek word stereoma, which means "a firm or solid structure." The Latin Vulgate (A.D. 382) used the Latin term "firmamentum" which also denotes solidness

and firmness. The King James translators coined the word "firmament" because there was no single word equivalent in English. Today, "firmament" is usually used poetically to mean sky, atmosphere, or heavens. In Modern Hebrew, raqia means sky or heavens. However, originally it probably meant something solid or firm that was spread out. (Translation by Joseph B. Lumpkin).

Talking about Genesis 1:6 (*And God said, Let there be a firmament in the midst of the waters, and let it divide the waters from the waters*), the theologian Martin Luther said: "*In Hebrew, the word 'Rakia' signifies 'a something extended,' from the verb 'Raka,' which means 'to unfold or expand'. And the heaven was formed by an extension of that original rude body of mist, just as the bladder of a hog is extended into a circular form when it is inflated.*"

Regardless of what we think is the meaning of the heavens, the heaven of heavens is for the Lord:

- "*Behold, the heaven and the heaven of heavens is the Lord's thy God, the earth also, with all that therein is*" (Deuteronomy 10:14).
- "*The heavens are thine, the earth also is thine: as for the world and the fulness thereof, thou hast founded them*" (Psalm 89:11).

God is bigger than the heaven of heavens. As he was preparing to build a house for God, Solomon knew that God is greater than any created things: "*But who is able to build him a house, seeing the heaven and heaven of heavens cannot contain him? Who am I then, that I should build him a house, save only to burn sacrifice before him?*" (2 Chronicles 2:6). After the building of the temple, Solomon repeated the same statement: "*But will God indeed dwell on the earth? Behold, the heaven and heaven of heavens cannot contain thee; how much less this house that I have built?*" (1 Kings 8:27). Knowing that as of today, Christians are the temples of the Holy Spirit, I better understood the "infinite" capability that human beings and especially born-again Christians have but that which they unfortunately ignore. I pray to God to help me use my gifts and abilities better for His glory!

Besides the creation narrative in Genesis, other books suggest that God is the one who made the heavens and the heaven of heavens:

- "*For all the gods of the people are idols: but the Lord made the **heavens**" (1 Chronicles 16:26).
- "*Thou, even thou, art Lord alone; thou hast made heaven, the **heaven of heavens**, with all their host, the earth, and all things that are therein, the seas, and all that is therein, and thou preservest them all; and the host of heaven worshippeth thee*" (Nehemiah 9:6).

This also implies that the heaven of heavens may not be where God was living or staying before creation. For He cannot be living in a place before creating it. God fills the whole universe and nothing is hidden before Him: "*Can any hide himself in secret places that I shall not see him? saith the Lord. Do not I fill **heaven** and earth? saith the Lord*" (Jeremiah 23:24). The psalmist alluded to the fact that the heavens of heavens need to praise God: "*Praise him, ye heavens of heavens, and ye waters that be above the heavens*" (Psalm 148:4). Many other verses add flesh to the description of the heaven(s):

- *"The Lord is in his holy temple, the Lord's throne is in **heaven**: his eyes behold, his eyelids try, the children of men"* (Psalm 11:4).
- *"The Lord looked down from **heaven** upon the children of men, to see if there were any that did understand, and seek God"* (Psalm 14:2).
- *"The **heavens** declare the glory of God; and the firmament sheweth his handywork"* (Psalm 19:1).
- *"By the word of the Lord were the **heavens** made; and all the host of them by the breath of his mouth"* (Psalm 33:6).
- *"Be thou exalted, O God, above the **heavens**: let thy glory be above all the earth"* (Psalm 57:11).
- *"Sing unto God, sing praises to his name: extol him that rideth upon the **heavens** by his name Jah, and rejoice before him"* (Psalm 68:4).
- *"The **heavens** are thine, the earth also is thine: as for the world and the fulness thereof, thou hast founded them"* (Psalm 89:11).
- *"The **heavens** declare his righteousness, and all the people see his glory"* (Psalm 97:6).
- *"But our God is in the **heavens**: he hath done whatsoever he hath pleased"* (Psalm 115:3).
- *"The heaven, even the **heavens**, are the Lord's: but the earth hath he given to the children of men"* (Psalm 115:16).
- *"My help cometh from the Lord, which made **heaven** and earth"* (Psalm 121:2).
- *"Unto thee lift I up mine eyes, O thou that dwellest in the **heavens**"* (Psalm 123:1).
- *"Our help is in the name of the Lord, who made **heaven** and earth"* (Psalm 124:8).
- *"Be thou exalted, O God, above the **heavens**: and thy glory above all the earth"* (Psalm 108:5).
- *"Whatsoever the Lord pleased, that did he in **heaven**, and in earth, in the seas, and all deep places"* (Psalm 135:6).
- *"To him that by wisdom made the **heavens**: for his mercy endureth forever"* (Psalm 136:5).
- *"If I ascend up into **heaven**, thou art there: if I make my bed in hell, behold, thou art there"* (Psalm 139:8).
- *"Bow thy **heavens**, O Lord, and come down: touch the mountains, and they shall smoke"* (Psalm 144:5).
- *"The Lord by wisdom hath founded the earth; by understanding hath he established the **heavens**"* (Proverbs 3:19).
- *"When he prepared the **heavens**, I [wisdom] was there: when he set a compass upon the face of the depth"* (Proverbs 8:27)
- *"I clothe the **heavens** with blackness, and I make sackcloth their covering"* (Isaiah 50:3).
- *"And I have put my words in thy mouth, and I have covered thee in the shadow of mine*

hand, that I may plant the **heavens**, and lay the foundations of the earth, and say unto Zion, Thou art my people" (Isaiah 51:16).

- "For as the **heavens** are higher than the earth, so are my ways higher than your ways, and my thoughts than your thoughts" (Isaiah 55:9).
- "For, behold, I create new **heavens** and a new earth: and the former shall not be remembered, nor come into mind" (Isaiah 65:17).
- "Thus saith the Lord, The **heaven** is my throne, and the earth is my footstool: where is the house that ye build unto me? and where is the place of my rest?" (Isaiah 66:1).

Although God lives in heaven and spirits have the ability to move very fast, not all spirits are allowed to live in heaven. For instance, it came to a point when, after disobeying God, Satan was cast out of heaven. For heaven is a place where God and some obedient or holy angels live. Hence, Satan was no longer qualified to live there after his fall:

- "How art thou fallen from **heaven**, O Lucifer, son of the morning! How art thou cut down to the ground, which didst weaken the nations!" (Isaiah 14:12).
- "For thou hast said in thine heart, I will ascend into **heaven**, I will exalt my throne above the stars of God: I will sit also upon the mount of the congregation, in the sides of the north" (Isaiah 14:13).

Jesus underlined the existence of a kingdom of heaven:

- "From that time Jesus began to preach, and to say, Repent: for the kingdom of **heaven** is at hand" (Matthew 4:17).
- "Blessed are the poor in spirit: for theirs is the kingdom of **heaven**" (Matthew 5:3).
- "Blessed are they which are persecuted for righteousness' sake: for theirs is the kingdom of **heaven**" (Matthew 5:10).

Jesus said that God the Father lives in heaven, and that only those who believe in God will go to heaven:

- "After this manner therefore pray ye: Our Father which art in **heaven**, Hallowed be thy name" (Matthew 6:9).
- "Not every one that saith unto me, Lord, Lord, shall enter into the kingdom of **heaven**; but he that doeth the will of my Father which is in heaven" (Matthew 7:21).

Like I explained in some chapters toward the end of this book, the day is coming when God will bring down a new heaven and a new earth, probably already existing somewhere in the universe: "And I saw a new heaven and a new earth: for the first **heaven** and the first earth were passed away; and there was no more sea" (Revelation 21:1).

Considering what I said above, to my understanding, no single contemporary scientific word related to celestial bodies or clusters of celestial bodies is enough to pinpoint what the Bible means by heavens or heavens of heavens. Sometimes, heaven is an atmosphere; other times, it refers to a celestial body like the moon or a star. Heavens sometimes allude to systems of celestial bodies, while the heaven of heavens alludes to a domain above all clusters of systems of bodies in the universe from which or above which God rules. Although other Jewish books also suggest a heavenly classification and ranking of the angels according to the territory they occupy, the terminology used does not relate to any conventional scientific jargon that most scientists can easily relate to.

Therefore, instead of trying to force any scientific term to explain heaven(s) and the heaven of heavens, I just take it as it is, knowing that they relate to physical and spiritual celestial bodies or systems of bodies that humankind may have a very difficult time decoding using modern scientific tools. As a result, in the chapters to come, I will not be using the term "heaven" or "heaven of heavens" a lot when I scientifically explain the formation of the universe, for they may not allow me to exactly address some questions people who are accustomed to modern scientific terminology may have.

Nevertheless, as far as the formation of the heavens and earth mentioned in Genesis 1:1 is concerned, the Bible gave some clues of what could have happened. Before I close this chapter, I would also like to mention what I came across from some Biblical commentators who have tried and done their best to explain the Biblical creation narrative in Genesis 1. Indeed, to understand what theologians have said about the meaning of the "heavens and earth" mentioned in Genesis 1:1, I searched and carefully investigated many commentaries of Genesis 1. While some of them are very heretical, others are well thought and very inspired. Some of these commentaries of Genesis 1 are hundreds of years old, yet the scientific community has not carefully looked at them to improve the scientific theories pertaining to the origin of the world.

I also came to realize that, by referring to the Hebrew meaning of the words used in Genesis 1, the creation narrative can be better understood. For instance, **Genesis 1:1 '*God in the beginning created the heavens and the earth*'** can be rendered in Hebrew as **"*Bereshith bara Elohim eth hashshamayim veeth haarets*"**. The Hebrew word "*bara*", rendered as "**created**", primarily expresses the start of the existence of a thing, from nonentity to entity. "***Bara***" is not the preservation or the formation of things that had previously existed, but it is creation. Some verses in Genesis 1 also suggest that "*bara*" denotes the formation of a body using a substance that was created. To put it another way, although it can also be used to signify the formation of a thing out of pre-existing matter (Genesis 1:21 and 27), the word bara is usually viewed as a production out of nothing. Creation is generally viewed as the making of a new thing or the bringing something out of nothing. To the strict definition of the term, creation must be the work of God only, for nobody else has the power to produce something out of nothing as God did in the beginning.

The Hebrew word "eth hashshamayim" is rendered in English as "heavens." However, the word "eth" generally means a particle or the substance of the thing. Hence, for centuries, some theologians have remarked that the Hebrew word "eth" means "the sum and substance of all things." In other words, the substance that God created in the beginning out of nothing is the first particle out of which the heavens and the earth were formed. As some theologians like Clarke Adam (1832)

CHAPTER 4: THEOLOGICAL EFFORTS TO DECODE THE MEANING OF THE "HEAVENS AND EARTH" IN THE CREATIVE NARRATIVE

put it, the word "eth hashshamayim veeth haarets" can be rendered as "the being or substance of the heaven and the substance of the earth". In other words, what Moses was recounting in Genesis 1:1 was not a finished or complete heaven and earth, as some people think, but their precursors, which, after going through some processes, yielded them. The precursor that I referred to in my writing as "turbulent prima materia" is the materials, the supplies, and the ingredients out of which God built the whole universe. Therefore, the sentence 'God in the beginning created the heavens and the earth' has been rendered by some theologians as "In the beginning, God created the substance of the heavens and the substance of the earth" (Clarke, Adam, 1832).

Therefore, to shorten a long story, the meaning of "In the beginning, God created the heavens and the earth" mentioned in the first verse of the Bible does not imply that God instantaneously created and formed everything in the world as they are today in the blink of an eye. A careful analysis of the creation narrative of the Bible and the scientific evidence suggested to me that God first created instantaneously the substance or the original matter that was used to form everything in the world indeed, but within the 6 days of creation, some processes took place after the initial creation out of nothing to shape the precursors of the bodies into what they are today. The challenging task I take up in the rest of this book is to explain those processes in light of the biblical stories and the scientific evidence.

Nathanael-Israel Israel: Known as the #1 International Authority that Truly Unlocked the
Secrets of the Turbulence that Shaped the Universe

CHAPTER 5

WHAT THE BIBLE SCIENTIFICALLY TEACHES ABOUT THE BEGINNING OF MOVEMENT AND ENERGY IN THE UNIVERSE THAT MOST PEOPLE IGNORE

5.1. Initiation of the movement of things

Everything in the universe is moving. Although the Bible did not specifically say why things were moving, it gave clues about what could have happened. According to Genesis 1, early in the beginning, the Spirit of God was moving on the surface of the waters. Just as He moved on the surface of the waters, some of which would become the Earth, the Holy Spirit would also have probably moved upon the surface of the precursors of other bodies being formed in the whole universe. The Spirit of God could have been what God used to move and position things in the universe. In other words, the movement of the Holy Spirit (at different positions of the precursors of the created things) could have been one of the factors, if not the main factor, that launched things into movement. Because of the depth and immensity of the data available on the motion of the Holy Spirit, I devoted a separate book to His role during and after creation. Indeed, in most references where the Holy Spirit or the Spirit of God is mentioned, things and even beings are usually moved. Likewise, when the Spirit of God was moving on the surface of the waters mentioned in Gen 1, He inculcated a movement to them. Consequently, the remote cause of the movement that celestial bodies and even particles have today, can be tracked back to the movement of the Holy Spirit. Like I explained in other chapters later, things in the universe today do not move at the same speed, in the same direction or sense, or the same way, because the way their precursors were shaped also imprinted changes to their characteristics. A comprehensive analysis of the direction and the sense of the motion of the planets, asteroids, and satellites in the Solar System allowed me to realize that the way they were launched into movement was coordinated.

To explain the action of the Holy Spirit to Nikodemus, for instance, God referred to wind, suggesting that the creation and the formation of the universe could have been very windy. The winds present in the vicinity of the Earth as of today could be much weaker than the ones that shook the precursors of the celestial bodies in the beginning. For, considering the energy, speed, size, and other characteristics of the bodies in the universe, the force that set them into movement must have been very powerful. Without setting my mind on the movement of the winds that prevailed during the formation of the universe, it may have been impossible for me to ever understand how the matter and system of bodies in the universe could have been formed.

During the formation of the universe, the Spirit of God could have manifested Himself in different ways: like a fire or like a rushing mighty wind, etc. The Spirit of God is often associated with a great, rushing wind. For instance, on the day of Pentecost, the Holy Spirit appeared to the disciples like a flame of fire and wind moving things and shaking the place (Acts 2:2-3; Acts 4:31). Similarly, in the Old Testament, there were days when God appeared to Moses like a fire in a bush, but the bush was not consumed. Many preachers referred to the Holy Spirit like fire. God is also called the "Divine Fire." He who talks about fire also implies energy. Talking about fire reminds me of the Azusa Street Revival (1906-1915), led by the African American preacher William J. Seymour (1870-1922), which birthed the Pentecostal and Charismatic movements at the beginning of the 20th century.

The same Spirit could have been the one that put the universe into motion in the beginning. At the time of Jesus Christ's death, a great earthquake also occurred in Jerusalem, proving the power and might of the Spirit of God, which, while leaving Jesus, shook the surrounding area. Even today, without their spirit, living things cannot move, for it is the spirit that moves and quickens things. That is why I cannot disagree with those who think that everything in nature has a spirit or an ability imparted to it by the substance created by God and which was transformed to form all things. Although some human movements are not controlled by the Spirit of God, some people are moved as the Spirit of God comes over, moves over, or moves in or upon them.

In the Old Testament, the Spirit of God moved upon some prophets and leaders. After a specific time, the Spirit used to leave them. For instance, that was the case for Othniel (Judges 3:10), Gideon (Judges 6:34), Samson (Judges 8:25), Saul (1 Sam 6:6), David (1 Sam 14:13), etc. The Spirit of God can also cause things to move in an unusual way and even make supernatural things occur, including causing a wild animal (e.g., a donkey) to speak (Numbers 24:2) or causing mountains to move (Zechariah 4:6). Under the Spirit of God, prophetic people can do supernatural things. However, it is important not to attribute all human motions to the Spirit of God. For instance, when some human beings are mad, they can become agitated and exhibit erratic behavior, not because the Spirit of God is moving them, but because of their own spirit or emotion. Likewise, those who believe in satanic rituals can also sometimes be moved by demonic forces which cause them to do things impossible for a normal, mere human being, suggesting the strong impact that spirits can have on the motion of matter. If time had allowed me to detail the behaviors of some demon-possessed people, many could be shocked at how much human beings and other things in nature are influenced by spirits. Just as a virus can enter a machine and affect its functioning, and just as a computer program can be run to execute codes, so also spirits are able to run encrypted programs (bad and good) written, encoded, inherited, and produced by and in human beings, other living things, and even nonliving things.

5.2. Why was the Spirit of God moving on the surface of the waters?

The motion of the Spirit of God on the surface of the water could have had many implications or purposes. For instance, knowing the power and energy of God, the movement of the Spirit of God upon the face of the waters must have destabilized the waters and generated turbulence, thereby affecting the formation of various structures within the fluids. The movement of the Spirit of God on the face of the waters was also to ensure that everything was being created, formed, shaped, positioned, and "empowered" according to God's plan, which also encoded characteristics, duties, and privileges to the creatures. As I delved into the scientific data, I came to realize that the Spirit of God should have moved at such a precise speed, in a precise way, and with a precise intensity so that the Earth, Sun, Moon, and other bodies in the Solar System could have been formed.

As I explained later, because of the link between the systems of celestial bodies in the universe, and because the movement of the Holy Spirit on the surface of the water in the vicinity of the precursor of the Earth was precise, it is important to know that the Holy Spirit must have also moved on the surface of many other precursors. Because God never changed and, subsequently, His Spirit could never have changed, His Spirit could have moved the same way but could have affected the precursors of the bodies differently according to some characteristics, including their size and the turbulence they underwent. To properly describe what happened during the genesis of the universe as the Spirit of God moved on the surface of the waters, it is crucial to understand that it was not just about the movement of the water (chemical labelled H_2O) or about the spontaneous fragmentation of the precursors of the bodies into daughter bodies without any process involved. On November 24, 2013, it appeared to me that, as the Spirit of God was moving on the surface of the waters during creation, He could have communicated or imparted some of His energy to them. For it may be difficult for things to come in direct contact with God or His Spirit without getting some impartation. As I detail the scientific data later, I will also expound on my thoughts.

5.3. Source of the first energy in the universe

On November 24, 2013, years before I ever thought about writing a book on the origin of the universe, the movement of the Spirit of God on the surface of the waters described in Genesis 1 caught my attention. As a prophetic believer, I personally believe that God created energy, but I did not know how He did it. Then, it came to my understanding that the movement of the Spirit of God over the created matter could have communicated to them a motion that, coupled with the mass of these bodies, could have yielded the first energy in the universe. Part of that energy could have been used to push the precursors of matter in different directions so they could be molded into various bodies (both small and large). Later, I will expound on these foundations I am laying here.

While I strongly believe that the Spirit of God could have been everywhere during the formation of the universe, I also believe that at the end of the formation of the world, His mission of creating the world may have "stopped", but He imparted into the bodies that were formed in such a way that, although everything created bears witness of God, the Spirit of God Himself is not in everything. In the Book of John, Jesus told His disciples (John 14) that He would ask the Father to send the Holy Spirit (which in Hebrew is called the Ruach ha-Kodesh) to them. This implies that the Holy Spirit was not in the world and that believers can receive Him by prayer when they ask God. The Holy Spirit is part of God. He has a creative power, and that is why anyone who believes in God becomes a new creature. The Holy Spirit can dwell only in those who believe in the word of God. That is why the Bible declares that unbelievers cannot be filled with the Holy Spirit. Similarly, this world cannot contain God nor His spirit.

Still, God can give life to anything He chooses. In fact, when Christ returns, the dead in Him (a.k.a. the believers in Yeshua) will rise from the dead by the power of the Holy Spirit, which gives life. In summary, wherever the Spirit of God is, there is true life. Because of that fact, and because of the holiness of God, the believers in God are the temple of the Holy Spirit (1 Corinth 3:16-17). The Holy Spirit in a believer grants him access to God the Father (Ephesians 2:18). He gives the believers victory over the flesh (Rom 8:2-4), tells them, and convinces them that they are children of God (Galatians 4:6). The Holy Spirit teaches the believers about God (John 16:14).

In contrast, the unbelievers "hurt" themselves and the Spirit of God by blaspheming Him (Matthew 12:3), resisting Him (Act 7:51), or insulting Him (Hebrews 10:29). When Christians sin or do evil things, they grieve the Holy Spirit (Ephesians 4:30-31, 1 Thessalonians 5:19). The requirements of the Holy Spirit could have been that, after creation, He left the Earth and probably many other places in the universe, leaving the omnipresence and many other attributes of God to continue His influence and presence. Today, and particularly after the death and resurrection of Jesus, the Holy Spirit dwells only in the believers who are seeking God and who are living a holy life.

There is a huge energy or speed associated with the Holy Spirit. When the Holy Spirit enters someone or something, He imparts part of that energy. That is why those who believe in the Holy Spirit and in the baptism of the Holy Spirit have testified many times that they feel heat over them. For instance, many charismatic preachers have testified many times that they sometimes feel heat in their bodies or all over their hands when they want to pray for the sick. In the Bible, Prophet Jeremiah also testified that the word of God is like a fire shut up and burning inside his bones and belly. Although some preachers sweat under the influence of the Holy Spirit, all kinds of sweating do not necessarily indicate a manifestation of the Spirit of God, for some sweating is caused by metabolism. The heat caused by the Spirit of God can differ from that caused by the Sun or by metabolic activity.

Seen from a holistic perspective, the energy in certain things and beings does not depend on their size, mass, or anything else physical. For instance, there was a time when people might have mistaken the dove that represented the Holy Spirit descending on Jesus during His baptism for a simple, ordinary pigeon. Yet, the Holy Spirit that was dwelling in that dove is able to do things that no human being could ever do. Likewise, it is important not to judge people and things using a mere physical analysis, for the encrypted invisible reality hidden behind the physical appearance of things is enormous. Therefore, most physical analyses that do not consider the spiritual world are very partial, limited, and unable to reach the truth. That is why some people have faced historic difficulties in mathematically solving some physical problems whose equations ignore supernatural parameters.

On May 29, 2018, as I began analyzing the scientific data from a turbulence perspective, I realized that the energy that moved things at the beginning of the world was so strong that it could have compressed anything below a certain threshold. God could have calibrated the movement of the Holy Spirit over the surface of the waters; otherwise, a completely different world would have been created because the turbulence generated could have been different. Because God never changed and cannot change, He had to create a huge universe instead of a small one that would have limited human beings, and also made Him look small. In other words, the energy in God and His plan for creation did not allow Him to create a small world that might not satisfy human curiosity or provide enough food for thought throughout the ages. To put it in another way, it is impossible for a mighty God to slow down His power and speed just to create a small world. Similarly, it is impossible for Him to spend many or even billions of years creating a world that He could create in a few days, or even a few seconds, if He had chosen to. Therefore, He chose to create a big world in a relatively short amount of time, 6 days, although He could have done it quicker if He wanted. Praise be to God for what He created. Unaware of this mystery, some human beings have been trying to comprehend the code He used to form the universe, not knowing that they would never fully comprehend it unless they believed in God, so that He would reveal to them some of the things He has done. As I addressed the formation of the universe, one of the challenges I also handled in the chapters to come is how the energy in the early universe was transferred to the bodies in the universe.

As I finish this segment on energy, I would like to mention that, regardless of whether it is seen from a physical perspective or a spiritual perspective, the world is filled with two types of energy: positive energy and negative energy. Although all forms of energy were created from a positive or good precursor, some of the energies in the world today have been corrupted and transformed. Satan, demons, and evil desires of human beings and other beings in the universe are the source of the corruption or change of the initial state of things from the best toward the worst before the end of this age occurs so that a new one (eternity) will begin or continue after a brief "interruption" of about 7000 years. For although human beings have a lifespan, there is an eternal component in their spirit and the temporary 7000 years that will pass (from the beginning to the end of the world) is like a brief moment in the eternity that has always existed before creation, and in which we are living now, and which will continue to exist after the "end" of the created world. I provided more details about this mystery in my book on the age of the universe.

As I introduce movement and energy in this chapter, it is also important to know that God can alter or reverse the movement of anything if He wishes. Similarly, those who believe in God can pray to Him to stop the motion of some celestial bodies. For instance, Joshua (the successor of Moses) prayed, and the Moon and the Sun stood still for hours: "*And the sun stood still, and the moon stayed, until the people had avenged themselves upon their enemies. Is not this written in the book of Jasher? So the sun stood still in the midst of heaven, and hasted not to go down about a whole day*" (Joshua 10:13). That story of Joshua is also recounted not only in the Bible but also in the Book of Jasher 88:63-65.

5.4. Birth of energy in the universe

The fact that kinetic energy is a matter in motion suggests that, at one point during the formation of the universe, something could have "pushed" the turbulent prima materia into motion so energy could be imparted to it. For, from the smallest things to the biggest ones, from the living things to the nonliving things, everything in the universe is moving, though some movements may not be easily seen with the naked eye or with the most advanced telescopes or microscopes. Although their movement cannot always be seen or felt, the constituents of matter are moving. Everything in the universe is a store of energy, which, under certain conditions, can be released to do work or produce heat, but also to manifest in other types of things or beings. Without an energetic input, most movements are impossible, implying that at the beginning of the formation of the universe, there was a mighty source of energy and/or a form of energy that was communicated or transferred to the initial matter. Otherwise, where did the universe find the energy required to ignite movements and form the various things it contains?

Along the years, it appeared to me that something happened to the turbulent prima materia for it to start moving, so that its daughter bodies could contain things that are moving although we cannot always see them with the naked eye. I showed that, as the bulk of the turbulent prima materia was broken into mega pieces under the influence of a violent explosion, energy was also relayed to those pieces. As the cascade of explosions or fragmentations of the precursors of the bodies went down hierarchically, energy and movement were also transferred to their daughter bodies accordingly. I also perceived that, the turbulent prima materia was rich in energy and moved from the top of the universe (i.e., the North "pole" of the universe) toward the bottom of the universe (i.e., the South "pole" of the universe) in the beginning of the universe.

The appearance of energy could have been so quick that separating the birth of the turbulent prima materia and the appearance of energy in it could not make any difference. Kinetic energy implies a mass in motion, of course; but a motion cannot start without the existence of a mass first, although both mass and motion can also appear at the same time. Therefore, for the sake of trying to explain things to a human mind who tends to see things linearly, it can also be all right to separate the appearance of the turbulent prima materia from its acquisition of energy, which implies a movement communicated to its massive body undergoing processes of split and gathering into daughter bodies. As the precursors of some clusters of matter were moving during the genesis of their bodies, some of their energy was used to tilt their orbital plane and rotational axis, while part of the energy was used to lock their constituents into compartments at distances that depended on environmental conditions in the early universe or at the beginning of the universe. At the chemical level, a significant portion of the energy used to form the particles is contained inside their constitutive particles. Because scientific laws suggest that energy cannot be created or destroyed, it appeared to me that the energy distributed in the clusters of bodies in the universe could have been born at an early stage of the formation of the universe. The energy communicated to a mother precursor of bodies as it was being molded was trapped into its daughter bodies. Part of the energy found in the bodies in the universe today could have been affected by their interactions with other bodies in their vicinity.

A difficult task I handled in this book, and more deeply in *"Turbulent Origin of the Universe"*, is how energy was transferred or split among the bodies in the universe. I found that the energy contained in matter could exceed what scientific instruments can measure. For, there are forms of energy that are not physical or chemical and yet real. Energy is not just a function of mass and speed only. The energy contained in chemical elements and in celestial bodies does not depend on their mass and speed only, but also on other characteristics related to how they were formed. The way energy is stored in bodies and the amount that can be released if the storage process can be "reversed" depend on many factors. Spiritual and mystic activities have also been occurring in some bodies in the universe and changing their energy.

The original matter in the universe could have been a kind of primal fire or flame from which all other bodies were born, just like sparks or flames can fly outward from a major flame and birth smaller flames or fires. Another way to express this is that, at certain stages of the development or differentiation of the precursors of the bodies, some of them could have been like a flaming fire or wind. This can also explain why all matter burns when a significant amount of heat is applied to it. For, fire was present at some of the stages of differentiation that some, if not all, precursors of the chemical particles passed through. The creation narrative in other Jewish scriptures (e.g., Book of Enoch) also alludes to the breaking of the precursors of matter in the early universe. In *Origin of the Spiritual World,* I provided more details.

'Science180 Academy' Success Strategy
SCIENCE180 MASTER CLASS

Hear the greatest scientific and philosophic lessons from top scientists, philosophers, thinkers, and public figures who have realized historic mistakes they made in life (concerning the origin of the universe, life, and chemicals), and that they corrected thanks to the historic discovery of Nathanael-Israel Israel, the world's first 180Scientist who founded Science180 and who is known as the one who truly decrypted the universe-origin for the first time. In their own words, these renowned personalities share with the world key lessons they have learned in life and how people can learn from their experiences to improve their lives, rather than repeating mistakes many still ignore at their own peril. To learn more, contact us at Science180.com/contact.

CHAPTER 6

CAN YOU SCIENTIFICALLY DEMONSTRATE THE BIBLICAL ACCOUNT OF THE UNIVERSE'S CREATION BY INCLUDING A MYSTERIOUS SCATTERING AND STRETCHING OF THE MOTHER OF ALL MATTER, DIFFERENT FROM THE BIG BANG?

6.1. Original mysterious scattering and stretching of the universe

The "original mysterious scattering", which I also called the "sudden burst asunder", or the breaking open, was an event that "ignited" the beginning of the spreading or stretching of the "turbulent prima materia," the initial matter in the universe. Soon after its appearance, the bulk of the turbulent prima materia was broken open by a violent event like a burst asunder accompanied by a huge noise as that of an "explosion." Although the internal pressure of the initial matter could have contributed to its "explosion" I felt like everything was under the control of the Spirit of God, which was moving along the surface of everything. I used the word "explosion" in quotes because it would not have been a mere explosion as we know today. Hence, I preferred to use the "original mysterious scattering" rather than "explosion."

The original mysterious scattering broke apart or divided the bulk of the turbulent prima materia into pieces or blocks of precursors of matter. It could have been possible that the turbulent prima materia broke apart because its surface tension was not strong enough to hold it together forever. The breaking apart of the turbulent prima materia exposed its content and "propelled" the broken pieces into many directions. In other words, after the first stage of the formation of the universe that was the sudden appearance of a bulk of initial matter, the second stage could have been the sudden breakup of that bulk of matter into fragments or pieces of bodies of different sizes. The energy communicated to each of the "broken pieces" of the turbulent prima materia could have propelled them into motion, and the bodies that were from them depended on the changes that their precursors could have gone through. The "broken pieces" of the initial matter started distancing themselves from the position they were before their mothers were broken. Their movement initiated the expansion of the early universe, which, even up until today, never ceased expanding. The expansion of the universe known today is a consequence of the force that propelled the broken pieces of the turbulent prima materia into motion, and spread them over the early universe. Today, because of the resistance of their crust, some celestial bodies such as planets, satellites, and asteroids may not be able to burst asunder despite the high pressure in their interior. However, when an escape is made in the crust, the internal pressure can release lava, which leads to volcanoes. In contrast, because they lack a crust, stars "explode" all the time, releasing some of their content (e.g., photons) into space. The surface of the Sun explodes all the time, but the Sun itself could no longer break apart because its precursor went through changes that "tied" its constituents together and prevented them from collapsing or breaking into pieces.

Some major daughters of the turbulent prima materia could have gone through additional "explosions" to yield smaller daughters. A cascade of "explosions" could have taken place until the "multi-broken pieces of the turbulent prima materia" could no longer explode or break apart. During these processes, changes were already occurring in the matter breaking apart. Some of these changes were caused by the turbulent instability occurring as things were moving. Some of the bulk of matter would later become precursors of clusters of galaxies. Some precursors of galaxies also "exploded" and yielded smaller clusters of stars, which seem to orbit a center of mass, which is not always a celestial body at their galactic center. Hence, some galaxies seem to have no celestial body at their center, unlike stellar systems, which have a star at their "center." Others are satellite galaxies that appear to orbit a larger galaxy. All of those clusters of galaxies and other celestial bodies in the universe are the product of how the precursors of bodies broke up and exploded during the cascade of events that shaped all the bodies in the universe. The daughter bodies born from each "explosive" event could have tended to form a system of bodies. The organization of the universe into galaxy clusters of various shapes and sizes is a consequence of the cascade of fragmentation and "coalescence" that the turbulent prima materia underwent to give rise to the bodies in the universe. Each defined cluster of galaxies, stars, or bodies can be traced back to a common precursor.

As I worked on the distribution of the celestial bodies, I also realized that the fragmentation of the precursors was not by means of a mere random explosion. As I explained later, specific laws related to fluid flow, turbulence, shearing, diffusion, and other complex phenomena in fluid mechanics and dynamics were involved so that the mother precursors of the bodies could be broken into primary bodies and secondary bodies accordingly. The creation narrative in other Jewish scriptures (e.g., Book of Enoch) also mentions an explosive event early during the formation of the universe. I think that the failure of previous theories on the formation of the universe may be the wrong interpretation of the origin of the explosion and how it shaped the bodies in the universe according to the design of God.

The creation and formation of the systems of matter or systems of bodies in the universe began at a place different from where they are today, but during their formation, they were relocated at the same time particles inside of them were also moved. Under the influence of the force that scattered them, the precursors of bodies also diffused through space. Besides the scattering force of the energy communicated to the precursors of bodies, their internal composition (e.g., density, viscosity) could have also affected their diffusion and, subsequently, the distance from the original position of their mother.

To sum it up, the universe did not start from a single point, as some theories suggest, but from a very deep, massive initial mass of matter. And if the expansion of the universe could be reversed and the movement of all bodies in the universe properly traced, it could be possible to estimate the direction or the location or region in space where the turbulent prima materia could have started its journey. Because of the expansion of the universe and the inability of current scientific equipment to properly measure the size, speed, and other characteristics of very remote celestial bodies, I did not spend much time trying to investigate the initial location of the turbulent prima materia. However, knowing that every particle and celestial body in the universe is moving, it seemed to me that the current location of things in the universe is not where they have always been since the beginning. For instance, although the Earth moves around the Sun and the Sun's movement across the sky seems constant, the Sun's location and that of the entire solar system are constantly changing. Similarly, although it is difficult or impossible to properly study the movement of all galaxies and clusters of stars in the universe, they are moving and changing places. Therefore, the position of the turbulent prima materia just at the beginning of the universe should be different than that of the current universe, for the whole universe is moving. In *"Origin of the Spiritual World"*, I provided more details on this matter. By the end of this book, I will explain why I think that the universe could have started at a position above the North Pole and not the south, west, or east.

The "broken pieces of the turbulent prima materia" could have gone through changes and processes, including breakup or fragmentation, squeezing or compression, and other stages of differentiation that I described in this book and more lengthily in *"Turbulent Origin of the Universe"*:

- Flow and instability of fluids
- Initiation and development of turbulence
- Splitting of fluid bodies and formation of the precursors of the bodies

- Transfer of energy
- Initiation of movement (e.g., revolution, rotation)
- Formation, splitting, and gathering together of fluid layers
- Acquisition and transfer of momentum
- Breakup, mixing, separation, stretching, tilting, or overturning of fluid bodies
- Spacing and sizing of bodies
- Birth and strengthening of forces that compressed matter
- Formation, squeezing, and wrapping of precursors of bodies
- And many more processes.

Many processes explaining the formation of the universe are also found in some biological systems, reactions, and pathways. For instance, in cell biology and development, the first or initial cells are undetermined, but as they go through stages of development, biochemical modifications, including cellular division, gene expression, transcriptomics, and epigenetics modifications, intervene to transfer the initially undetermined cells into determined cells, which are not always able to reverse back to other kinds of cells, including the undetermined cells. Similarly, not only was the turbulent prima materia undetermined, but it was also transformed into specific daughter bodies, most of which were unable to reverse back to their precursors. Although some chemical elements can be transformed into others, in general, the processes of transforming particles of one kind to another do not usually occur naturally on a large scale today. The energy and processes involved in shaping things in the universe have been costly, and it is not easy to reverse natural laws or things. Even at the level of human beings with a will, it is not easy to change human nature. Else, using their free will, human beings could have been more easily changing their minds and nature to adapt to new requirements or to become whatever or whosoever they wanted to be. In other words, natural laws are very strongly forged and are not easily breakable using part of the energy that formed them.

Unlike living organisms, whose bodies usually start with one egg or cell, which then undergoes many mitoses before differentiation begins, in the case of the universe, things started with a large mass of a single kind of matter that filled the early universe. Just as undetermined cells started splitting into different parts with different functions, so did the bulk of the turbulent prima materia also go through fragmentations simultaneously associated with modifications of the characteristics of their daughter bodies.

6.2. Processes involved in the formation of the universe according to the Bible

To implement the design that He did for the world, God used His divine wisdom (Proverbs 8:22-29). Before I scientifically start talking about how the creation and formation were done, I need to draw your attention to some processes that the Bible suggests occurred during the formation of the universe. These processes are dispersed in many books in the Bible. Although for thousands of years all of these processes could have been well known to those who read the Bible, the main difficulty or challenge in understanding them has been how to reconcile them to fit into a non-contradictory story that must also consider the massive scientific data and the commentaries that Bible scholars have done and that believers have held onto throughout the ages.

Processes suggested by Genesis 1

- Creation of an initial matter out of nothing by the spoken word of God
- Movement of the spirit on the waters
- Induction of turbulence
- Separation of waters from waters (fluid separation)
- Formation of the atmosphere
- Formation of the crust
- Formation of stars and satellites
- Establishment of rotational movement
- Gathering together or a collection of layers of fluids

Processes suggested by Job

- Measurement of the foundation of the Earth
- Tilting of the celestial bodies by God
- Passage from gas-like and liquid-like to solid-like state of the precursors of bodies as they gained their shape
- Inclination of the precursors of bodies, suggesting a progression in the change of state of the precursor as their matter was being converted into their current shape

Processes suggested by John 1

- The Word can create physical things or can be molded into physical things and beings
- Existence of a relationship between darkness and light

Processes suggested by Psalms

- God made the world using His hand

Processes suggested by the epistles of Apostle Paul (e.g. Ephesians, Colossians)
- God is not in the heavens but above the heavens–in *"Origin of the Spiritual World,"* I explained how many heavens exist and on which one God resides.

Processes suggested by Apostle Peter
- The Earth will be destroyed by fire

Processes suggested by Apostle John in the book of Revelation
- The structures of the universe may be like a tree bearing fruits
- The Earth will be destroyed one day and replaced by another one

Processes suggested by the Pseudepigrapha (the lost and rejected books)
Although I did not detail them in this book, the lost and rejected scriptures revealed some processes not mentioned or detailed in the Bible. These processes include for instance the formation of water, angels, the shaping or cutting out of stars out of fire of the precursors, the positioning of the celestial bodies at a specific distance, etc. In *"Origin of the Spiritual World,"* I addressed all of these processes on the formation of the universe.

More details about the secrets that the Bible revealed about creation can be found in chapters 15 and 16.

CHAPTER 7

WHY YOU DON'T HAVE TO EMBRACE EVOLUTIONISM AND TRASH THE BIBLICAL ACCOUNT OF CREATION AS IF THE BIBLE LIED ABOUT HOW THE SPOKEN WORD OF GOD CREATED THE FIRST MATTER?

More and more people are easily and conveniently embracing evolutionism and trashing the Biblical account of creation as if the Bible lied about how the spoken word of God created the first matter. I am here to tell the whole world that this trend is extremely dangerous and that there is a way to fix it accurately.

7.1. Creation of the initial matter in the universe out of nothing

Besides the Book of Genesis and the Gospel of John, whose first chapter is devoted to creation, other references in the Bible prove that God created the world with His spoken word: "*The universe was created through a spoken word of God, so that what is seen did not come into being out of existing phenomena*" (Hebrews 11:3). After creating the universe out of nothing, God used His Spirit (Holy Ghost or Holy Spirit, called Ruach ha-Kodesh in Hebrew) to move over the things He created, to shape and confer to them different missions. Consequently, if anything found in the world could be unfolded or reversed back to its initial reactants as it was just at the beginning, what would be found would be the word coming out of the mouth of God. In other words, everything in the world is a product or daughter of the reactions induced into space (space itself was created) by the spoken word of God. For some reason, I could not see why I should not believe that, before creation, God never spoke a word until the first time He did, and the world was created. But before speaking, God had already planned everything and caused the reactions to His speech to follow a creative pattern that, in the end, would be used to redeem some of His creatures. In other words, just as a human being can be quiet, thinking, formulating a plan and strategizing in his mind and waiting for the appropriate time before speaking, so also to a much higher and supernatural extent, far beyond what a mere human being can understand, God took eternity to plan, measure things, and elaborate strategies, interactions, and missions from the final outcome. He wished to accomplish it and knew how to begin it, so His plan never failed. He "rehearsed" that plan and ensured everything was perfect, "waited" (in an era where there was no current time) until the proper time came, and He just spoke the word, and the world instantaneously came into existence. Some prophets in the Bible who reported on the voice of God said that it is like the voice of many waters, or like thunder. Because of the crucial role that the word of God played in the beginning, God also used His son Jesus (Yeshua) to be the means by which people would be saved, and those who refuse to believe in the word of God will never see God but will spend their eternity in darkness, far from Him.

Although most Christians believe that God created the world, a big and challenging question that, to my knowledge, no scientist before me has scientifically answered is how God formed the universe after creating the first type of matter out of nothing. That is a thought-provoking question I have handled in a language that even unbelievers can comprehend and accept.

It has been proven that sounds propagate like a wave. Spoken words can produce waves. Many things that move can sometimes destabilize certain things in the space around them, and even cause some waves to form, although these waves cannot always be seen. When human beings speak, they can generate waves. Across space and the universe, many things move and behave like waves. Even light behaves like particles and waves. Scientific data proved that subatomic particles (i.e., particles smaller than atoms) behave like or show some characteristics of waves, implying that every matter in the universe is made of a certain kind of waves. These examples suggest that, across the universe, something could have happened at the beginning that caused waves to exist and persist. In other words, the wave-like characteristics of many things in the universe point to a common denominator or foundation of everything in the universe: an initial wave, which can be related to the initial word of God. The problem with human beings is that they are unable to create or form matter with their own waves. For the waves they produce are limited by the conditions existing on Earth, for the environmental limits on Earth and the weakness of the waves that human beings can produce cannot allow these waves to behave as God's waves did in the beginning. However, in the beginning, when nothing was created or formed yet, the waves of God's words could have behaved differently, and, under the right power or instructions, they could have been transformed into the created matter. Nevertheless, some waves produced by man-made equipment or machines today can be felt like objects. Some waves are so powerful that they can affect the movement of particles in space or produce sounds that can affect things around them. Most sounds move more slowly than light, and the faster a sound is emitted, the higher the likelihood that it may move things.

Some anointed men and women of God have called into existence miraculous things through prophetic declarations using the word of their mouth. Some prophets who have undeniable experience with creative miracles testified that in the spiritual world, words are objects. For a word spoken by a spiritual master has the power to materialize into things in the physical world. That is why blessings and curses can follow spoken words. In other words, God and people can speak a word to bless or transmit a blessing; so can evil people. Across the globe, parents tend to bless their children by speaking a word over them. Many people also teach the power of positivity. In other words, people believe that positive words can have a positive effect on people's lives, while negative words can be damaging. Yet, many people do not believe that God used His word to create the whole world.

Spiritual spells, incantations, bewitchments, invocations, charms, curses, and enchantments are proofs of the possibility of a word to materialize into physical things. For instance, magicians, witches, sorcerers, and enchanters know well that spoken words can be used through enchantments and other spiritual rituals to curse people, inflict pain, cause disease, kill, fake things, negatively change people's behavior and destiny, and even cause some supernatural reactions to materialize into the physical realms. Yet, most of those people do not believe that God Almighty spoke the words and that the whole world was created, and that the same God is able to deliver people using the same word or by bringing to pass the word spoken by His servants on Earth. Those who believe in positivity are not completely wrong. Even those who do not believe in God have testified to the power of the word that people speak over themselves. People become more like what they say and confess than what they do not. If this is true for a mere human being, how come some people cannot believe that God created everything with His Word? How can we believe in the power of our words to create our world and yet dare to refuse to believe in the power of the word of God Almighty to create the whole world?

It is not by chance that the tongue is so powerful. It is one of the most energetic and consequential organs. God used His tongue to utter words that created the world. Consequently, He gave power to the human tongue to utter words that can create things good and bad. Sadly, most people use their tongues not to create or build good or positive things, but to destroy, curse, bring people down, and craft evil things filled with negative energy! In other words, the human tongue has the power to build and destroy, while God used His tongue to create good things. Let's be careful of what we say to ourselves and to others using our tongues. It is better to stay silent than to say negative, destructive, and hurtful things with our tongues. There is power in the tongue! Why not use your tongue to praise and worship God? In addition to speaking, even as of today, some people can "create" or make certain things happen just by deeply thinking about them. For instance, using a type of thinking called "rayoni," some human beings can engage in such deep thinking that it becomes a vision because God takes control of it, and the things associated with it will surely come to pass. This calls for a need for all to better control what we think and say to ourselves and to others. Likewise, although some people are gifted with the power of creative miracles, no human being can do what God did to create the world!

Did you ever ask yourself, why matter listens to and obeys spoken words? In other words, if positive words can affect human beings, it means that matter can listen to spoken words. Things listen when human beings speak because they were made with the spoken word of God, and they keep that ability inside of them. If matter had never listened and responded to the spoken word during creation, it would not be able to do so today.

Some people think the Hebrew language was the one God spoke during creation and in the Garden of Eden. At the same time, some Christians who believe in the gift of tongues, or who speak in tongues, think that those tongues were the original language. Some people also believe that all creatures speak and hear, and that they hear and understand human beings when the latter speak in tongues. Hence, some preachers recommend speaking in tongues to better communicate with nature, including giving instructions to the creatures. Although I do not know how to prove that, I cannot disagree with those people. Indeed, until the days of Noah, meaning before the flood, all of the inhabitants of the earth spoke one language: "*And the whole earth was of one language, and of one speech*" (Genesis 11:1), which, like I just said, some people believed was the Hebrew language. Likewise, it could have been possible that the language that Adam and Eve spoke in the Garden of Eden was Hebrew. It was after the flood that God confounded human language: "*Therefore is the name of it called Babel; because the Lord did there confound the language of all the earth: and from thence did the Lord scatter them abroad upon the face of all the earth*" (Genesis 11:9).

The Latin phrase "ex nihilo," meaning "out of nothing", is usually used to describe the creation of God out of nothing. When I use the term "formation," I usually mean a process that uses the created material to mold something new or different. The Bible also uses the verbs "make," "form," "fashion", to express a process of molding the initial matter into something else. The term creation "ex materia" is sometimes used to mean the formation of things out of some pre-existent material. Unfortunately, some people seem not to believe that the creative narrative in Genesis 1 is a creation "ex nihilo", but they think it is a creation "ex materia", therefore mistakenly believing that God did not create things out of nothing. What a mistake! Just as the word was used to create the universe, so also the word of God that people believe in makes them become new creatures. For the word of God never lost His creative power! Just as the creation of the universe was miraculous, so also the creation of believers into a new creation is marvelous and supernatural.

I would like to say a few words about the speed of the word of God. Indeed, just as how the sound that comes out of the mouth of a human being propagates like waves, so also the sound that came out of the mouth of God at the beginning could

Nathanael-Israel Israel: Author of "Science180 Accurate Scientific Proof of God" have propagated like a mighty wave. At the beginning, there was no universe yet, meaning that the space in which the sound of God's voice propagated could have also been created by the word of God as well. The sound emitted by God could have been very huge, powerful, and fast, even faster than anything existing in the universe today. To be able to create the universe in 6 days, the word of God must have acted very fast, else He could not have traveled throughout the entire universe and done what He did in such a short amount of time. Knowing the omnipotence of God, nothing is impossible for Him and His word could have traveled the whole world in less than the blink of an eye, and created matter everywhere. The movement of the word of God and His impact on the created things can explain why everything in nature is moving, and nothing is static. According to science, sound travels much more slowly than light, and it can take millions of years for light to cross some galaxies. According to science, the Milky Way (one of the countless galaxies in the universe) has a diameter of more than 100,000 light-years. Although this statistic can be biased, I believe that the word of God must have traveled at a speed higher than that of light. Because the word of God is the same, He created the same original matter everywhere before later transforming it to form different things according to their location and mission.

During the formation of the universe, God designed a process so that the bodies formed do not move at the same speed as the initial matter He created, otherwise, human beings for instance could not be able to use some of the things on Earth today, for they could have been moving faster than them. For instance, if everything on earth today were moving as fast as the word of God, human beings may not be able to use them for they may be unreachable, unless human beings could also be moving at such speeds. In contrast, unlike what the scientific community claims, I think certain particles and bodies in the universe could be moving at a speed faster than that of light.

On July 23, 2015, more than 2 years before I discovered turbulence in the data I studied on celestial bodies, as I was thinking about the waves produced by the word of God, it appeared to me that if I could study the speed, mass, and position of particles and celestial bodies with respect to their primary body, I could notice that the speeds of all particles and bodies in space are related to a common source or reference that can be traced back to the common waves caused by the word of God! Because the spoken word of God created the world and everything in it, I believe that the word of God is the source of the wave-like nature of things in the universe. I do not mean that every wave in the universe is caused by the word of God, for as of today, many artificial waves are caused by anthropogenic activities. However, the source of natural waves can be traced back to the origin when God spoke His word. As the created things were molded to form other things, waves still continue to exist, and the footprint of the wave related to the word of God was left in everything formed. The movement of celestial bodies (e.g., light, planets, stars, etc.) has been causing other waves in space. Because celestial bodies can behave differently according to the seasons and time, their waves can vary consequently.

Therefore, waves across space can vary according to the position of the bodies producing them. Scientists have claimed the discovery of a common wavelength present in every part of the universe, but they fail to connect it to what could have generated the initial waves throughout the universe. God is able to act upon the entire universe. In the next segments, I will explain the characteristics of the initial matter and how it was molded into everything in the universe today.

7.2. Characteristics of the initial matter in the universe

In my book called *"Turbulent Origin of the Universe"*, I established that, in the beginning, a certain kind of matter, which I call the "turbulent prima materia", or the "turbulent original particle", or the "turbulent original substance", or the "foundational matter", nothing quite as similar to any matter known today, meaning its characteristics are different from that of any matter known today, was mysteriously created by the word of God out of nothing, ..., and through very complex, dynamic, and turbulent processes, that turbulent prima materia was progressively molded into all types of matter, bodies, systems or clusters of matter, and bodies in the universe today. The Bible does not provide much information about the nature of the initial particle that constitutes the heavens and the Earth God created out of nothing in the first verse of the Biblical narrative of creation in Genesis 1.

Due to the complexity of what could have happened in the beginning of the world and the simultaneous reactions that could have occurred at various locations in the early universe, it is difficult to describe chronologically and exhaustively in a linear format, list all the processes or steps that took place from the beginning of the world to its current state. Therefore, do not confuse my efforts to tell a story in this book, chapter by chapter, with the exact chronological succession of events that occurred from the beginning of the world until now.

Just as during the development of a living organism, an egg can go through steps of differentiation to become different types of cells, tissues, organs, apparatuses, or systems, which are all connected, so also during the formation of the universe, the "turbulent prima materia" went through complex processes to birth all kinds of matter and systems of bodies in the world today. The original matter in the universe was like an "undetermined" particle, which had the potential to become anything just like at the beginning of the development of a living thing, the first cells have the potential to become anything and are later specialized into specific cells because of unique changes in the expression of their DNA. Similarly, the initial matter in the universe had a program that produced different types of matter depending on the conditions it was subjected to. The difficult task is to explain the machinery of the processes involved in the shaping and distribution of that "turbulent prima materia" into the bodies present in the universe today.

The initial matter in the universe could have occupied a very huge portion of space which can be labeled as very deep and wide. For instance, the massive volume of the precursor of the Earth and of other celestial bodies on the first day of creation could explain why the Bible referred to some precursors as "the deep" as the movement of the Holy Spirit over the waters was addressed (Genesis 1:1-2). Because the limits of the universe are still unknown to human beings, I did not try to estimate the size of the space occupied by the initial matter. As the mother of all matters, the "turbulent prima materia" was formless, leaving the process that shaped it after its appearance to be the reason for the diversity of matter in the world today. From this point forward, I used the term "daughter" to refer to bodies born from a mother's body.

As of today, matter is generally found in 4 states: solid, liquid, gas, and plasma. I also understand that most schools teach only 3 types of matter: solids, liquids, and gases. Considering the similarities and differences between these states of matter, I perceived that the status of the "turbulent prima materia" was none of the 4 states of matter known today. In other words, the initial state of the original matter had to go through complex changes and processes so that each of the current 4 states could be formed.

Although it is impossible to properly describe the initial matter out of which everything in the universe was formed, I felt like at one point, the "turbulent prima materia" could have looked like a flash of lightning, a flame of a burning fire, a hot fire, or like the "magma" or "lava" from a volcano. I used all of these terms to try to describe the initial matter because when I saw some bodies in the universe, such as the Sun and the magma or lava that comes out of the Earth's interior, I felt like they were witnesses of what the initial matter could have been made of. I am not saying that the turbulent prima materia was magma or lava or a light—for each of these types of matter was formed after the original matter was molded—but I am trying to use current facts to illustrate what happened in the beginning. For instance, according to the creative story of Genesis 1, light was not the first matter that God created; for light was made using a matter that was initially created. Because no particle in the early universe was left unchanged by the processes that shaped the world, nothing in the universe today is an exact photocopy or saving of the initial matter in the beginning. The laws of physics must have changed as the universe was being formed and like I demonstrated in *"Turbulent Origin of the Universe"*, natural laws are not the same everywhere in the universe nor the same throughout the steps or stages of changes that the universe went through from its beginning until today. This is similar to how the relationship or interactions between parents and their children change during the stages of life: pregnancy, birth, childhood, teenage age, adulthood, marriage age, procreation of the children and birth of grandchildren. In other words, how family members interact when a woman is pregnant is different from how they do after the children are born, and become teenagers, adolescents, pubertal, adults, and ready to be married, then married and having their own children, until the parents are very old and about to die. Likewise, the relationship between family members varies from one family to another. Likewise, the laws of nature are not the same everywhere like most scientific theories claimed.

Finally, as a historical reminder, I would like to say that, on January 1, 2014, meaning more than 10 years before I published this book, it appeared to me that the original matter that God created was obedient to God, else God could not have used it to form other bodies. The flexibility of the initial matter is a key to its ability to be the foundation of everything. Although some pseudepigraphic books suggest that some groups of stars are bound up because they did not obey God, meaning that some matter could have "sinned", I decided not to delve into those in this book. Here, "sin" may not mean the same thing as what human beings do, but an inability to go through certain changes and finish the process of formation as certain bodies did.

Another Book by Nathanael-Israel Israel:
TURBULENT ORIGIN OF CHEMICAL PARTICLES

FIND ALL THE RELIABLE, CONVINCING, SCIENTIFIC
ANSWERS YOU NEED TO SUCCESSFULLY DECODE THE
ORIGIN OF CHEMICAL PARTICLES SAFELY

Where did all elementary particles and composite particles, including atoms,
molecules, minerals, and rocks, come from? What are the fundamental
factors, the machinery, and the generic processes that define their
formation and properties? What was the nature of their precursors at the
beginning of the universe and what underlying processes shaped or molded
them into the chemicals we know today? What was the primary cause of the
abundance and diversity of chemicals in the celestial bodies in the universe?
What is the accurate link between the formation of chemical particles and
the formation of galaxies, stars, planets, asteroids, and satellites? What light
can the origin of chemicals shed on the real cause and meaning of gravity
and the other so-called fundamental forces in nature? How does the
formation of the chemical particles fit into the big picture of the formation
of the universe?

After studying these questions for more than 12 years, Dr. Nathanael-
Israel Israel discovered that the proper understanding of the origin of
chemical particles is a very challenging but profitable task that requires
original, scientific, mathematical, and philosophic efforts beyond the
current state of modern science—until recently. The solution for all of
these puzzling problems: *"Turbulent Origin of Chemical Particles"*, the
straightforward and trustworthy book that will help you to quickly, cheaply,
easily, and efficiently navigate everything you need to know to finally solve
the hard problems about the origin, the formation, and the functioning of
all chemical particles. Whether you are a chemist, a biochemist, any other
scientist, or an engineer, if you have a reasonable background in chemistry
but ignore how to scientifically demonstrate the origin of all chemical
particles, this marvelous book is for you!

Amazingly packed with eye-popping analysis, fantastic graphs, tables,
and the historic formula that broke the universe-origin code, *"Turbulent
Origin of Chemical Particles"* will:

- Make it easier than ever for you to properly understand, decrypt, and articulate the real origin of natural chemical particles in the universe, therefore freeing you from false and boring explanations of the origin of all matters, and embrace the proven theory that opens doors to unparallel opportunities
- Professionally teach you how to transform the true knowledge of the origin of chemical particles into insights that significantly add value to your life in less time, and successfully establish you as a symbol of freedom, power, creativity, and originality in your field of expertise
- Fire you up to become the best version of you, and to cause positive changes to your initiatives that will profit you nonstop
- Discover thrilling illustrations and unconventional explanations of the formation of all matter in the universe, written in a simple language that brings humankind much closer to the complete deciphering of the mysteries at the very heart of chemistry, and open the way to a future of technology, innovation, discoveries, and breakthroughs
- Equip you to bypass technical knowledge that restricts non-experts from accessing the origin-related secrets contained in the massive scientific data, and get to the bottom of origin-related mysteries regardless of your background so you can empower yourself to leave unforgettable marks in your field of expertise
- Learn more at Science180.com/chemical

With *"Turbulent Origin of Chemical Particles"*, the accurate decryption and understanding of the formation of chemicals has never been profitable and easy. Hence, this great book is THE ultimate how-to guide great people wanting to correctly decode the origin of the chemicals and positively transform their lives. Get this celebrated book today. Don't wait!

Known as the nonconformist, rule-breaker, and accurate demonstrator of the universe's origin, **Dr. Nathanael-Israel Israel** is the founder of Science180, the one-stop for answering the most crucial universe and life's origin questions. He has had the honor to be acknowledged as the fearless universe-origin decryption trailblazer. Learn more at Israel120.com.

CHAPTER 8

CAN WE SCIENTIFICALLY PROVE THAT FLUIDS (INCLUDING WATERS) EXISTED AND WERE DIVIDED FROM OTHERS DURING THE FORMATION OF THE UNIVERSE LIKE MOSES SAID?

8.1. Theories of the fragmentation or breakup of matter

According to the creation narrative of Genesis 1, bodies of water were separated from one another, while others were collected together during the formation of the universe. In addition to scientific data, this story suggests that a fragmentation of fluids occurred during the formation of the universe. Because I will talk a lot about processes of fluid fragmentation throughout the rest of this book, I felt like before I go any deeper into the details, I need to summarize what is known about that subject. Scientists have found that fluid fragmentation is a ubiquitous phenomenon that can be classified into three groups:

- Aggregation
- Maximum entropy principle and random breakups
- Sequential cascades of breakups

Aggregation is defined as a process by which an *"ensemble of initially small elementary particles form clusters of increasing average size as they collide and merge, and the evolution is towards ever larger sizes"* (Eggers and Villermaux, 2008). I used to use the term "aggregation" to explain the process by which precursors of celestial bodies were clustered into entities originally separated, but I finally realized that a collision of precursors was not what really happened during the formation of the universe. Other studies (including those of the previous authors) also suggested that collision-based fluid fragmentation is rare in nature. To avoid confusion and complication in my writing, and to distance my work from errors in some existing theories, I decided that instead of saying "aggregate," I chose to use the terms "gather together", "assemble together", or "collect together". Consequently, instead of using the term "aggregation," I adopted the terms "gathering together", "assembling together", "collecting together", or "collection together" to express the process by which the precursors of bodies originally separated, dispersed, or isolated precursors of bodies were collected together or gathered together into more unified entities or systems of entities under the influence of turbulence. The data I studied also suggested that the splitting and gathering of the precursors of bodies, or of systems of bodies, into their daughter bodies were not separated or isolated in time. As the precursors of the bodies were splitting, their constitutive or internal components were also being gathered together. Therefore, to translate the joint process of splitting and gathering together the precursors of the bodies, I invented the term "split-gathering together" or "split-gathering" in short. Because during the formation of the universe, all processes of breakup and gathering together of the precursors of bodies were turbulent, I invented the term "turbulent breakup", "turbulent gathering together", or "turbulent split-gathering" to emphasize the role of turbulence. I also realized that the processes by which matter was clustered into more condensed bodies or systems of bodies were not by means of what is called "accretion" as postulated in some theories, such as the Big Bang.

Let's now talk about the second group of fragmentation: the maximum entropy principle and random breakups. Indeed, the maximum entropy principle and random breakups approach considers the random splitting of an initial volume into various disjointed elements, all in one step (Eggers and Villermaux, 2008). Like I explained throughout this book, the fragmentation of the precursors of the bodies in the universe was not random and did not happen all in one step or at one time, yet it happened quickly.

Finally, the process of aggregation is inverse to the process of a sequential cascade of breakups, which implies size reduction. It is a process where a mother's body gives rise to daughter bodies, which themselves break into smaller bodies, producing ever smaller sizes, and so on and so forth.

In 1941, the Russian mathematician Kolmogorov (1941) was the first to model sequential cascades of breakups. He was inspired by ore grinding, a process where the size of brittle solid particles is repeatedly reduced. Kolmogorov later applied that model to turbulence (Kolmogorov, 1962; Monin and Yaglom, 1975), and despite its errors, his turbulence model has been applauded by many in that field of research, not because it was perfect, but because no better replacement had been found for more than 60 years. In each of the variants of the cascade process, a drop (of liquid) of initial volume breaks, after n steps of the cascade, into a family of drops.

The three classes of fragmentation models were compared with experimentally measured and statistically converged drop-size distributions (Eggers and Villermaux, 2008). It was found that: "*maximum entropy principle and aggregation scenarios were found to be far from the truth: nature does not aggregate nor split liquid volumes at random*". No perfect model exists yet to properly explain the fragmentation of bodies in nature, and more or less at the beginning of the universe. Nevertheless, sequential cascades of breakups seemed to be closer to the reality found in nature.

After about 7 years of investigating the subatomic particles, atoms, molecules, minerals, rocks, celestial bodies, and clusters of galaxies, I understood that the formation of the universe balanced a sequential cascade of breakups with a process of gathering the constituents of the daughter bodies following the breakups. At this point, I would like to briefly review what is known about fluid breakup today and how it may apply to the story of the universe's formation as presented in Genesis 1.

8.2. Application of fluid breakup processes to the formation of the universe

Fluid breakup, or thread breakup, is the process by which a single mass of fluid breaks into several smaller fluid masses. The breakup of jets is most often driven by surface tension (Eggers and Villermaux, 2008). According to Wikipedia (2019), fluid breakup is characterized by the elongation of a fluid mass forming thin, thread-like regions between larger nodules of fluid. The thread-like regions continue to thin until they break, forming individual droplets of fluid. Similar to taking silly putty or chewing gum in your hand and stretching it by pulling with both hands. You will have a clump in each hand while the area between each hand stretches and becomes thinner.

To explain the breakup process without diluting it too much, I will use some scientific terms here. If you do not understand all of them, do not worry; for, like a doctor who breaks down a medical language into terms his patients can understand, I will do my best to explain myself in a language everyone can comprehend. Indeed, thread breakup occurs where a fluid in a vacuum form a free surface with surface energy. If more surface area is present than the minimum required to contain the volume of fluid, the system has an excess of surface energy. A system not at the minimum energy state will attempt to rearrange so as to move toward a lower energy state, leading to the breakup of the fluid into smaller masses to minimize the system's surface energy by reducing its surface area. The exact outcome of the thread breakup depends on the surface tension, viscosity, density, and diameter of the thread undergoing breakup. According to Wikipedia (2019), the breakup of a fluid thread or jet usually begins with the development of small perturbations (e.g., vibrations of the fluid container or non-uniformity in the shear stress) on the free surface of the fluid. The wavelength of the perturbation is the critical parameter in determining whether a given fluid thread will break up into smaller fluid masses. The fluid viscosity can also affect how rapidly a given perturbation grows or decays over time. Changes in the internal pressure of a fluid thread can be induced by capillary pressure as the free surface of the thread deforms. Capillary pressure can also depend on the curvature of the interface at a given location at the surface. This implies that the fluid pressure depends on the two radii of curvature that shape the surface of the fluid. Within the thinned area of a fluid thread undergoing breakup, the first radius of curvature is smaller than the radius of curvature in the thickened area, leading to a pressure gradient that would force liquid from the thinned areas to the thickened areas. However, the second radius of curvature remains important to the breakup process. For some perturbation wavelengths, the effect of the second radius of curvature can overcome the pressure effect of the first radius of curvature, inducing a larger pressure in the thickened regions than in the thinned regions. This would push fluid back toward the thinned regions and tend to return the thread to its original, undisturbed shape. However, for other perturbation wavelengths, the capillary pressure induced by the second radius of curvature will reinforce that of the first radius of curvature. This will drive fluid from the thinned to the thickened regions and further promote thread breakup. I know this paragraph may not make much sense to you, but, don't worry; as you keep reading, you will understand what needs to be understood.

I perceived that the fragmentation and gathering together of the precursors of matters in the universe are 2 phenomena that must be simultaneously considered to fully explain the formation of the universe. For one did not happen without the other. In other words, regardless of their size, the precursors of the bodies in the universe were not gathered together without being previously fragmented. I do not mean that the gathering of bodies must always follow fragmentation. It seems impossible or unreasonable for things or bodies in the universe to be gathered together or assembled if, at one point, their components were not initially fragmented into separate entities. There are levels of fragmentation and of gathering together, and things were not fragmented and collected together in the same way.

Because under certain conditions, a single fluid mass can break into many smaller fluid masses, it appeared to me that the precursors of the celestial bodies could have gone through "cascades" of breakups that yielded the bodies in the universe. Some of those breakups could have started with the elongation of the fluid mass of the precursors of bodies. As the precursors of the bodies were moving under the influence of turbulence, some could have been amassed into larger clusters of matter that were still going through changes. Some could have been fluid clusters (of various sizes), which could have been elongated, then formed thin, thread-like regions between larger fluid clusters. These thread-like regions could have continued to thin until they broke from the ligament that originally attached them to their mother precursor, forming individual fluid masses with various sizes and characteristics. Here, the size of the fluid threads could have been as small as possible, but also as astronomically big as possible. For, anything is possible, and processes that formed the smaller kinds of matter also formed the big ones, but with diverse calibrations. The size and location of the precursors could have been some of the factors that defined the machinery that molded and shaped their destiny. In other words, the environment where the precursors of bodies landed during the fragmentation and clustering of the initial matter is one of the factors that defined what they became.

As fluid threads of the precursors of the bodies were being formed, some broke up because of factors including their free surface energy, which was also going through changes. The precursors whose surface area was more than the minimum required to contain the volume of their fluids could have had a surplus of surface energy. The precursors, which were not at their minimum energy state, would have attempted to rearrange in order to diminish their energy state by reducing their surface area. This rearrangement could have led to the breakup of their fluid into smaller masses. While all of those processes were going on, other characteristics acquired by the precursors of the bodies affected the ways their constitutive bodies were rearranged. Some of the factors I will explain later include viscosity, density, spatial localization, and composition of the thread. Each of those factors could have affected how rapidly a perturbated fluid could have grown or decayed over time. For instance, during the formation of the bodies in the universe, the surface of the precursors could have also been perturbed because of their movement through space and also because of other factors such as the turbulence that was occurring inside of them. According to the perturbation that acted upon them, some precursors could have produced a thread that contributed to fragmenting them into other bodies. In contrast, the thread of precursors whose perturbation was weak might have stayed in a state that did not allow them to be fragmented.

On December 26, 2017, soon after I realized that the scientific data that I was dealing with concerning the formation of the universe was about turbulence, I also noticed that, the breakup of the precursors of the celestial bodies into their daughter bodies (daughter precursors) was very quick. According to the Bible, by the end of the 4th day of creation, some celestial body could have been formed. On the second day of creation, the Bible says that waters were split or divided from waters, suggesting that waters or liquids broke up into others or were separated from others. This may also be referring to the breakup of the precursor of the Earth-Moon system from the precursor of other bodies in the solar system, and/or the breakup of the precursor of the Moon from the precursor of the Earth, and/or the breakup of the precursors of celestial bodies near the Earth-Moon system. Given that the formation of the Earth was finalized by the 3rd day, while that of the Sun and the Moon was on the 4th day, it is reasonable that from the 1st day to the 4th day, the breakup of the precursors of many bodies occurred in the universe. The fast speed of the breakup may also be explained by the high speed of the precursors of the celestial bodies. These precursors were moving fast because they were pushed into motion by a very powerful energetic source. For instance, the high speed of light suggested to me that, during the formation of the universe, certain precursors moved very fast. For if the precursor of light was not moved or launched at least as fast as that of light (final product of that precursor), it may be impossible for light itself to move so fast. The breakup time of the precursors of the celestial bodies may have depended on their size and many other factors, which I will explain later.

'Science180 Academy' Success Strategy:
SCIENCE180 ACADEMY PROGRAMS

Owned by Science180, Science180 Academy is a training, speaking, consulting, and mentoring program specialized in everything universe-origin, life-origin, chemicals-origin, and anything at the intersection of reason and faith, or science and religion.

Science180 Academy deals with different subjects according to the needs of its members or target groups. When people register for Science180 Academy, they must choose the program(s) they want to focus on so their training can be properly personalized accordingly. This is similar to how people register for a university and take classes in a specific department matching their needs!

Science180's breakthroughs are so complex and dense that it is not realistic or good to try to explain all in just one academy, else people will be overwhelmed, disinterested, and confused by the plethora of data to handle. In other words, Science180 Academy offers a wide range of origin-related training in various domains strategically designed to allow people to choose the most suitable for their needs so that, regardless of their background or field of expertise, people can equip themselves, align their mindset, and improve lives today and forever using the accurate explanation of the origin of the universe, of life, and of chemicals. Science180 Academy curriculum is based on 12 years of deep unconventional research that culminated with the publication of many much-admired books on the formation of the universe and its content (see www.Science180.com/books).

The content of each Science180 Academy is strategically crafted by Dr. Nathanael-Israel Israel (who is acknowledged as the internationally-acclaimed world's authority in origin-related issues) to suit scientists and nonscientists, religious and nonreligious people, leaders (as well as experts), and followers so they can fully decode the proofs of the formation of the universe, of life, and of chemicals they have been wanting to demonstrate or grasp.

The current programs of Science180 Academy are the following:

1. SCIENCE180 ACADEMY OF COSMOLOGY (Designed for all scientists who want to scientifically study cosmology, the science of the origin and fate of the universe)

2. SCIENCE180 ACADEMY OF TURBULENCE (This is a perfect fit for scientists and other experts interested in studying abiotic turbulence).

3. SCIENCE180 ACADEMY OF LIFE SCIENCES (Tailored to those who want to study biotic turbulence)

4. SCIENCE180 ACADEMY OF CHEMISTRY (Designed for chemists, biochemists, scientists, and other educated people who want to understand the origin of chemical particles)

5. SCIENCE180 ACADEMY FOR LAYPEOPLE OR THE GENERAL PUBLIC (Very fit for any layperson or "less" educated people who wants to learn (in a simple language) deep insights that even those who went to university for years were unable to decrypt by themselves...

6. SCIENCE180 ACADEMY FOR CHILDREN (This academy breaks down origin key topics into language that children can fully understand). This is the only Science180 Academy that your whole family will like and enjoy together, and it will set children on the path of success by accurately showing them early in life the formation of the universe, and how to detect errors in theories or stories that would misguide them as they grow up.

7. SCIENCE180 ACADEMY OF THE PSEUDEPIGRAPHA AND SPIRITUAL WORLD (Only one ancient blueprint has the reliable power to help you to accurately decrypt the spiritual origin and history of everything in the universe. If you are a believer and want to delve into the prophetic, angelic, and higher order of knowledge based on the spiritual world, then this Science180 Academy is for you. This program is suitable for those who took at least "Science180 Academy of Creationism".

8. SCIENCE180 ACADEMY OF CREATIONISM (Science180 Creationism is a scientific theory spearheaded by the groundbreaking discoveries of Nathanael-Israel Israel, that scientifically explained the origin of the universe, life, and chemicals using turbulence and that mathematically reconciled science and the Biblical account of creation for the first time in history. Science180 is different from all existing creationist theories known before 2025. Science180 Creationism reconciled science with the Biblical account of creation, including scientifically proving that the Earth was formed on Day 3, while the Moon and the Sun were formed on Day 4 of creation!).

9. SCIENCE180 ACADEMY FOR FREETHINKERS & ALL ANTI-CREATIONISTS (This Science180 Academy is designed for evolutionists, anti-creationists, and all other types of unbelievers seeking to rationally explore and understand alternative arguments for the creation or formation or origin of the universe, life, and chemicals from a fresh scientific perspective).

10. SCIENCE180 ACADEMY OF LEADERSHIP-(Also called "Science180 Academy for Leaders", this program will enlighten leaders of organizations on how to solve their people problems, process problems, and profit problems related to the origin of the universe, of life, and chemicals according to their domain of expertise). With "Science180 Academy of Leadership", leaders will gain new insights so they can cast new visions and avoid focusing on screwed-up processes, products, and services related to universe-origin initiatives that need to be fixed, faced, or dealt with. Science180 Academy of Leadership will also equip leaders to address process problems related to inefficiency, gaps, missed opportunities, wasted time and efforts, too many steps, bureaucracy, useless layers between organization and customers concerning the innovation, research methodology, research, product development, ...

11. SCIENCE180 ACADEMY FOR GOVERNMENTAL AGENCIES (Do you want to know how and why most nations and governments are wasting millions of dollars on universe-origin and life-origin research they don't need ... and how to avoid it?) Indeed, for most developed nations, and even for some underdeveloped countries, universe-origin projects can cost billions of US dollars and other expensive things that cannot be afforded without sacrificing crucial priorities. Even in developed countries, the impact and the return on investment of the space research are subjects of intense political and economic debates. What if your nation or institution can reduce wasteful spending on universe-origin research and life-origin research, as well as your dependency on wrong theories on the origin of the universe? and life?

12. OTHER SCIENCE180 ACADEMIES: If you did not relate to any of the Science180 Academies mentioned above, but you are still interested in learning something specific about the origin of the universe, life, and chemicals that better fits your needs, please visit Science180Academy.com to contact us so we can discuss them with you.

CHAPTER 9

WHY YOU DON'T HAVE TO REJECT THE LITERALISM OF GENESIS TO CONVINCE PEOPLE ABOUT THE TURBULENCE THAT REIGNED DURING THE CREATION OF THE UNIVERSE?

9.1. Beginning of turbulence in the universe

In simple terms, an instability is a phenomenon or process that destabilizes things. For instance, when a fluid is hardly shaken, it can be destabilized. Likewise, as the "broken pieces of the turbulent prima materia" started moving and stretching under the influence of the "original mysterious scattering" (that I described in a previous chapter), the whole universe was destabilized, and major turbulence broke out.

When fluids are heavily destabilized, their constituents rearrange differently into "pockets" or compartments whose composition and size can depend on the intensity of the instability. For instance, when cream is poured into a coffee, an unstable mixture can be seen. Similarly, when water or any other liquid is put in a bowl and then thrown into the air, depending on the intensity used and the way it is thrown, that liquid not only can leave the bowl but also, as it starts moving in the air, it can be scattered into clusters of liquid of different sizes, speeds, shapes, etc. There is a way to destabilize, scatter, and cause a fluid to yield daughter clusters that can gain rotational and orbital movement. In the same manner, there is a way to use a single force to simultaneously launch things into an orbital and rotational movement.

Considering all matters in chemistry, biology, and physics, and considering the thousands of pages I wrote about their characteristics, it appeared to me that the origin and movement of all matters can be traced back to a major instability that occurred during the early stage of the formation of the universe. Scientific evidence suggests today that the universe is expanding. At the same time, it is accepted that the universe is maintained together by some natural forces such as the so-called fundamental forces in nature (e.g. gravity). For the universe to keep expanding while things in it still exist and generally "maintain" their identity, the initial instability at the beginning could have been able to fragment the initial matter while also allowing its daughter bodies to gather together according to some processes, which, in the end, allowed the formation of a complex and dynamic universe filled with moving things.

Under the initial instability, the turbulent prima materia started changing its state, moving, and mixing with one another, leading to the formation of vortical structures. In short, a vortex is a swirling fluid mass. The initial matter could have moved in a direction compatible with the force that destabilized it. During their movement, the unstable fluids began to split and gather into precursors of bodies of all sizes. These precursors of bodies became the precursors of other bodies and so on and so forth until the smallest bodies or clusters of bodies were formed. The movement of some fluids was like a flow. The instability of these fluid flows could have played a huge role in the organization, size, location, and structure of subatomic particles, atoms, molecules, planetary systems, stellar systems, galaxy clusters, and the universe as a whole. From the smallest types of matter to the galactic systems passing by planetary systems, stellar systems, bodies in the universe were assembled into different groups based on how their precursors were moved around by the events that "disturbed" the turbulent prima materia.

As the precursors of matters were flowing, fluid layers were formed, and stacked one on top of the other and moved at different speeds. Turbulence started in them, and patterns or rearrangements of their matter emerged. During my first 7 years investigating the origin of the universe and its contents, I realized that the physics underlying many phenomena in nature is related to turbulence. Turbulence is considered as "the last great unsolved problem of classical physics". In fact, it is one of the main unsolved problems in science and one that some scientists, including many Nobel Prize winners, have wrongly taught that God cannot even explain.

No clear definition exists for turbulence, a subject that has been explored by scientists for more than 500 years. Turbulence is usually defined by the experts in that field as a state of fluid motion characterized by seemingly random and chaotic multi-dimensional vorticity. Turbulence is usually referred to as chaos, a state of disorder, randomness, trouble, or a disordered motion of a "crowd." But as for me, turbulence is not chaos but a high level of order that people have failed to perceive. There is a certain order that people cannot understand and call disorder. As I was working and reflecting on the scientific data available on the universe, I realized that they are the astronomical expression of what some scientists are looking for in the lab and complaining that they are not able to collect or comprehend because of the limited size of their lab experiments and the limits imposed on human scientific spectra of understanding. To put it another way, I perceived that the lack of advancement in the field of turbulence and in any field related to the origin of the universe is not due to a lack of data, but the fact that people did not know the meaning of the data already collected and the thought process they need to get the most out of it. I also realized that all of the kinds of turbulence in the universe can be linked to the initial turbulence that I call the "Mother of all turbulences," the strongest and largest turbulence of all time. Before that turbulence, no other turbulence existed and since it occurred so the universe could be formed, no other turbulence as strong has occurred yet although many turbulences are still happening in the universe on many weaker scales.

The intensity of the turbulence at the beginning is not always the same from one precursor of matter to the other. Not only did the turbulence contributed to the separation of the precursors into different bodies, but it also contributed to imparting different characteristics to them according to their location. The pattern of turbulence at the origin of the universe had structure on all imaginable scales, from the invisible to the tiniest to the largest. Each of those scales of turbulence went through different development stages and led to the diversity of the bodies and systems of bodies in the universe. The kinetic energy that the precursor of some daughter bodies received from their mothers allowed them to overcome the eventual resistance that the viscous force of their constituents could have had. The fluids in the daughter bodies could have been subject to different intensities of turbulence, leading to the formation of different bodies accordingly. The nature of the daughter bodies depended on many factors, including their location. The richness of energy of the initial matter can explain why all matter contains energy and can be seen as a "modified version of assembled energetic particles." In other words, the data I analyzed (which I detailed in my book *Turbulent Origin of the Universe*) suggested that, during the formation of the universe, turbulence occurred on different scales, with different intensities, in different positions, and led to the formation of bodies in the universe. From the smallest microscopic scales to the largest astronomical ones, the scales of the turbulence that I studied in this book can be classified as follows:

- Invisible or spiritual scale that no naked eye or scientific equipment will ever see
- Microscopic scale (that of atoms, molecules, minerals, rocks),
- Celestial bodies scale (that of stars, planets, asteroids, and satellites)
- Stellar system scale (scales of stars)
- Galactic scale.

However, for the sake of time and space, I will not detail all these turbulence scales in this book, but those who are interested in learning more can refer to *Turbulent Origin of the Universe* and *From Science to Bible's Conclusions*.

9.2. Reference to turbulence in the Bible

On a cold midnight of December 14, 2020, as I was standing outside my house (for a family member to come home), as I have been doing for the last 9 months since Covid-19 hit the globe, my mind was drawn toward some previous scientific work that could have allowed people many centuries ago, even before the time of Isaac Newton, to understand that the creation narrative of the first 4 days in Genesis 1 is filled with turbulence details. I later understood that some of the scientific studies done centuries ago, such as those by Copernicus, Kepler, Galileo, Isaac Newton, etc., were enough for them and others to explain the turbulent events that Moses summarized in Genesis 1, but they and other authors before and after them failed to do so because they focused on other models, and also, they did not think that the story of Moses was really a chronological account of events emphasizing the formation of the Earth, the Moon, the Sun, and other bodies that God created.

On December 14, 2020 (a day I fasted and prayed), I realized for the first time that the tohu bohu mentioned in Genesis 1:2 referred to turbulence. Indeed, the *"without form and void"* in Genesis 1:2 (**The earth was without form and void)** was originally rendered from the terms **"tohu and bohu"**. The Hebrew epithets "tohu" and "bohu" designate anything empty, confused, vain, and worth "nothing." These epithets carry the idea of confusion and disorder. Hence, by referring to "tohu bohu" as "chaos" or a state of confusion, some people think that chaos existed in the beginning. Because chaos is also related to turbulence, it is fair to say that, during the formation of the universe, mighty turbulence took place under the influence of the Spirit of God.

God could have created (bara) the basic particle or substance of all things as an impressive and deep mass of matter, which in its original state was without arrangement, or distinction of parts. The British Methodist theologian and biblical scholar Adam Clarke (1762-1832) referred to this state as a *"vast collection of indescribably confused materials, of nameless entities strangely mixed, a crude and indigested state of the original substance, … that God spent 6 days assimilating, assorting, and arranging into all of the bodies in the universe"* (Clarke, 1832).

The "spirit of God" (Ruach Elohim) mentioned in Genesis 1:2, and which moved upon the face of the waters, has been thought by some theologians to be a *"violent wind"*, an *"elementary fire"*, the *"Holy Spirit"* (see John 3:8 and Acts 2:2 according to which 'a *mighty rushing wind on the day of Pentecost, filled the house where the disciples were sitting'*). The word *"moved"* used in Genesis 1:2 came from the Hebrew word *"merachepheth,"* *which means "brooding over"*, which some theologians explained by using the example of the '*motion made by the hen while either hatching her eggs or fostering her young"*, therefore evoking the idea of incubation, or hatching an egg (Clarke, 1832). The previous author reported that some theologians have interpreted the Hebrew word *"merachepheth"* *(meaning 'moved')* as a movement that communicated a vital or prolific principle to the waters upon which the "Ruach *Elohim"* (the Spirit of God) was moving. This interpretation may have been what caused some people in antiquity to believe that the *"world was generated from an egg."* As I showed in my writings, although this myth concerning the "egg" is false, it nevertheless carries a significant concept of the processes pertaining to the split-gathering of the precursors that I detailed in *"Turbulent Origin of the Universe,"* the scientific version of my book on the origin of the universe. In other words, just as an egg goes through processes to yield a chick (that grows and matures), so also the precursors of the whole world (was like an "egg") went through processes that formed the universe not from a physical egg, but through very complex procedures that the ancients (even those who lived before Jesus Christ) cannot grasp other than by using the term "egg".

While referring to Genesis 1:2 (*And the Spirit of God moved upon the face of the waters*), the German professor of theology and pivotal figure in the Protestant Reformation, after which the word "Lutheranism" (a branch of Protestantism) was coined, Martin Luther (1483-1546) said:

> *"And as a hen sits upon her eggs that she may hatch her young, thus warming her eggs and as it were infusing into them animation, so the Holy Ghost brooded as it were on the waters; that He might infuse life into these elementary substances which were afterwards to be animated and garnished. For the office of the Holy Spirit is to give life [...] Moses meant that in the beginning of the world, the heavens and the earth were created by God out of nothing; but created in a rude shapeless mass, not formed and beautified as they now are"* (Luther, 1904).

The powerful Spirit of God moving at the surface of the waters must have caused turbulence astronomically bigger than the turbulence caused by any category 5 hurricane or any tsunami! By the time I wrote this chapter, I had finished writing *"Turbulent Origin of the Universe,"* the scientific version of this book, and I had already discovered the role that turbulence played in the formation of the universe. I had already extensively addressed turbulence in *"Turbulent Origin of the Universe"* indeed, but until December 14, 2020, it never appeared to me that the expression "tohu bohu" or "chaos" used in Genesis 1 is an allusion to turbulence. The quote *"And God saw that it was good"* was not pronounced by God at the end of Day 2, as done for all the other 6 days of creation, because the precursor of the Earth (which was one of the main foci of the story told in Day 2) was going through turbulence that was shaping the precursor of the Earth as well as the precursor of the Moon and the Sun. As I later explained, the Moon and the Earth came from a common precursor, and because the formation of the Moon was finalized on the 4th day, its precursor was still undergoing turbulence, as was the case for the Earth's precursor on the 2nd day. By this time, the precursor of the Earth-Moon system could have been split into the precursor of the Earth and the precursor of the Moon. Some people may think that the Bible did not provide many details about the precursor of the Moon, but as I will explain later in this book, a careful analysis and understanding of the events associated with turbulence and the details provided for the formation of the Earth gave a lot of clues as to what could have happened to the precursor of the Moon and the precursors of many other celestial bodies including the Sun, whose formation is mentioned in the events of Day 4.

As a result of the turbulence on Day 2, God had to wait for Day 3 for the formation of the Earth to be finished and plants formed before God said, "*And God saw that it was good*". On Day 1, God also said "*And God saw that it [light] was good*" to express His satisfaction with the light He made. The formation of light did not take much time. God did not say that everything on Day 1 was good because, as mentioned in Genesis 1:2, turbulence was still ongoing with the precursors of some bodies. The light could not have been the only thing formed on Day 1. For, the formation of some bodies (chemical particles and even celestial bodies in other systems) could have been really finished on Day 1. God chose to focus the story of the light because God's main interest for the story of Day 1 was to inform human beings about things in the vicinity of the Earth.

The absence of the "*And God saw that it was good*" at the end of Day 2 is one of the things that caused some theologians and Bible commentators to assume that the Devil fell on Day 2, supporting that assumption using some verses in Ezekiel and Isaiah that point to chaos when the Devil was ejected from heaven. As for me, that supposition of the fall of the Devil on Day 2 of the creation narrative is a terrible mistake, for by the end of Day 6, the world that God created was found perfect by God himself, and that is why He rested on Day 7. I used to think that the fall of Satan may have happened no sooner than Day 7, and the chaos that accompanied the fall of Satan was different than that expressed in Genesis 1. Then, as I was reading the pseudepigraphic scriptures, I noticed that some Jewish literature stated that Satan fell even before God rested on Day 7. I elaborated on the exact date of the fall of Satan in *"Origin of the Spiritual World."*

The turbulence of a fluid associated with the formation of the universe cannot be assimilated to the turbulence connected to the fall of Satan and some disobedient angels. I think that the inability of people to interpret the story of Day 2 regarding "chaos" is an allusion to the turbulence that the precursor of the Earth (and probably the precursors of other bodies as well) went through, and that the Earth mentioned in Genesis 1:1 is a precursor of the Earth, not the "complete" and well-shaped Earth as described by the end of Day 3, is part of the misunderstanding that caused some theorists (even Christian scholars) to assume that the Earth was already completely formed by the end of Day 1 and that the chaos mentioned on Day 2 is about the fall of Satan and his angels. I wondered why those who claim that Satan fell in Genesis 1:2 cannot understand that this was even before the formation of the Earth was completed on Day 3, meaning that this cannot happen, for other biblical verses also mentioned the life of Satan in the Garden of Eden before his fall. How can those verses be explained if on Day 1 of creation, people think that Satan already fell while the Garden of Eden was not even planted on Earth yet but on Day 6?

The difficult part of explaining the origin of the universe is how turbulence affected the formation and characteristics of all bodies in the universe. The mentioning of turbulence in Genesis 1:2 does not mean that it was only on Day 1 that turbulence existed in the universe, but it was born on Day 1 and continued until today (although on different scales) in some environments across the universe. To be precise, turbulence in the universe could have started with the instability introduced by the Spirit of God by the moment He started moving over the first matter created by God out of nothing. I will provide ample details later.

Another Book by Nathanael-Israel Israel:
TURBULENT ORIGIN OF LIFE

THE ONLY ACCURATE FORMULA TO SCIENTIFICALLY EXPLAIN THE FORMATION OF ALL FORMS OF LIFE QUICKLY

Every human being will benefit from understanding the real origin of life. But the problem is that most efforts to explain the origin of life are complex, inaccurate, confusing, partisan, and complicated, therefore, creating serious challenges to those who are eager to scientifically decrypt where all forms of life came from. Most people want an accurate, simple, straightforward, nonpartisan life-origin book that is free of jargon and difficult concepts known only to experts. This elegant scientific book breaks down the technicality of the origin of life in a language that even nonscientists can easily comprehend. It is a trustworthy book that will help you to quickly, cheaply, easily, and efficiently navigate everything you need to know to finally decode and solve the puzzling problems about the origin of life, while also giving you a crash course on the universe's origin.

Unlike any book you have ever read on the origin of life, this historic masterpiece (which distills complex scientific data into simple explanations that make sense) is the starting point for any smart person wanting to rationally understand the formation of all living things. By the time you finish reading *"Turbulent Origin of Life"*, you will discover the following:

- Why in spite of the massive amount of scientific data collected on living things, scientists have misunderstood the formation of life until now, and then uncover in a simple language the one thing that was needed to accurately crack the code of life, but that scientists have missed and that has been causing them headaches, overwhelm, and burnout
- Step-by-step pathway to decode the origin of life and get the power, freedom, and boldness to take advantage of the opportunities that an accurate understanding of the origin of life creates (Science180.com/life)
- The high connection between the code of the universe formation and the process by which life on Earth was formed so you can become a fulfilled thought leader in your field of expertise

- Tools to stand as a lightning bolt that electrifies those who are still struggling to understand the formation of all forms of life in the universe
- Strategies to push the boundaries of human abilities to properly understand what is perceived as un-understandable, mysterious, supernatural, unimaginable, impossible, and unthinkable that hold people back
- Scientific approach to holistically detect, correct, and remove all misinformation, ambiguity, and misleading claims and theories surrounding the origin of life

Whether you are a scientist or a layperson, a believer, or a skeptic, you cannot afford to ignore the greater, better, faster, simpler, cheaper, easier, and accurate formula unlocked in this important book that successfully decoded the origin of life. Get *"Turbulence Origin of Life"* today and change lives! Don't wait!

Dr. Nathanael-Israel Israel is the Father of Science180 Cosmology, and the Founder of Science180 Academy. He is fortunate to be known as the source of unconventional wisdom and knowledge that helps people accurately crack the code of the formation of the universe, of life, and of chemicals. Get some resources by visiting his personal website at Israel120.com.

CHAPTER 10

READ THIS TO DISCOVER A HISTORIC FINDING OF THE 21ST CENTURY ABOUT THE CASCADES OF TURBULENT SPLIT-GATHERING THAT OCCURRED EARLY IN THE UNIVERSE

10.1. Generations of turbulent split-gathering of precursors of bodies in the universe

How did the bodies in the universe get to their current place or position? How did the first matter that God created reorganize or split to birth the various things in the universe? In this chapter, I will explain the process that was involved in this split and gathering of the precursors of the bodies in the universe. Indeed, everything in the universe was formed using the initial matter or substance that God created with His word. This process is what some people call "creation out of nothing." In other words, some of the things created in the universe and apostrophized in Genesis 1:1 had to go through changes, modifications, and molding before becoming what they are today. The main challenge in any theory of the universe's formation is explaining how the things we see today were molded (spontaneously or progressively) from an initial matter that served as the precursor to all things.

The diversity and the timing of the events associated with those processes are the foundation of all or most errors, misunderstandings, and diversity of the theories on the formation of the universe. The pathway the universe's precursors took to wind up in its current configuration is very difficult to understand using current scientific methods alone. Throughout *"Turbulent Origin of the Universe"*, I explained, using science, how the universe and everything (living and nonliving) in it were formed. A correct understanding of the origin of the universe and how it functions will also open new avenues for science, including physics beyond the standard model, which obviously contains flaws.

On December 25, 2013, as the world was celebrating Christmas, it appeared to me that the Earth mentioned in Gen 1:1 must have been a precursor of the Earth, for other verses in the same chapter suggest that the Earth went through some changes before yielding the Earth as we know it today. Without having many details about the processes involved, I felt since 2013 that some reactions occurred before the dry land and wetland appeared on Earth. Likewise, I knew since those days that the heavens mentioned in Gen 1:1 must also include the precursors of other bodies in the universe. It took me about 7 years before coming across some commentaries on Genesis 1, which also suggest the notion of precursors of bodies. The processes of the formation of the universe I described in this book and others explain the creation and/or the formation of the initial matter, darkness, light, galaxies, stars, planets, asteroids, satellites, microscopic matters (subatomic particles, chemical elements, molecules, minerals, and rocks), and all living things, including angelic beings.

In the context of the formation of the universe, I defined a generation of split-gathering of the precursors of matter as their division into groups of precursors that reflect not necessarily their date of birth, but mostly the stage, hierarchical position inside other systems of bodies, or along the chains of breakups, and the nature of most of their daughter bodies. Clusters of bodies of the same generation of breakup could more likely have been born after the turbulent primary material went through similar cascades of breakups. Even so, as I explained later, due to phenomena such as intermittency (the uneven distribution of size, energy, and dissipation in physical spaces), some smaller bodies formed as larger clusters were split, without their precursors undergoing many cascades of breakups. For instance, the precursors of galaxies belong to the same generation of breakup, while the precursors of stars belong to another generation, and the precursors of planets belong to another generation, while the precursors of satellites belong to another one. Because some precursors could have been formed quickly, what I labeled as "cascades" of breakups could have occurred very rapidly and in some instances spontaneously or simultaneously in such a way that it could be better to view the "cascades" of breakup as a succession of fragmentations that varied more in space than in time. As some bigger bodies were forming, smaller ones inside of them were also going through their own process of fragmentation and gathering together. Under other circumstances, some mothers must have literally gone through some fragmentations before their daughters' bodies could have experienced their own. For instance, unlike human beings, whose children have a developmental growth younger than their mother's, meaning that mothers give birth to children, who must grow up to become adults before birthing their own children (therefore giving grandchildren to their mothers), the fragmentation and gathering together of matter and clusters of bodies were different. All daughter bodies in the universe could not have waited for their mothers to finish their development before going through their own. When the mothers of some celestial bodies were going through their development stages, some of their daughters could have already birthed their own children, which could even be growing and taking good care of their own children before their grandmothers finished their formation or development. Considering the organization, characteristics, and the current classification of the clusters of bodies and clusters of matter in the universe, I concluded that about 9 generations of breakup and generations of gathering together of precursors could have occurred during the formation of the universe:

- 1st generation of turbulent split-gathering, leading to the cascades of precursors of galaxy clusters and globular clusters
- 2nd generation of turbulent split-gathering leading to the precursors of stellar systems
- 3rd generation of turbulent split-gathering, causing the precursors of stellar systems to form the precursors of primary stars and the precursors of the bodies orbiting them (e.g. planetary systems, planets without satellites, asteroid systems, and asteroids without satellites)
- 4th generation of turbulent split-gathering, causing the precursors of planetary systems to birth the precursors of primary planets and their satellites, while the precursors of asteroid systems birthed the precursor of their primary asteroids and their satellites
- 5th generation of turbulent split-gathering, causing the precursors of satellite systems to form the precursors of individual satellites and of rings
- 6th generation of split-gathering, leading to the precursors of minerals, mineraloids, and rocks
- 7th generation of turbulent split-gathering, leading to the precursors of atoms and their clustering into molecules and chemical compounds
- 8th generation of turbulent split-gathering, leading to the formation of the precursors of subatomic particles
- 9th generation of turbulent split-gathering, leading to the precursors of the smallest particles that will never be scientifically discovered

For more details about each generation of turbulent split-gathering, please consult the scientific book called *"Turbulent Origin of the Universe,"* in which I also explained the phenomenon of intermittence, according to which smaller bodies are usually found between bigger or larger bodies. Due to the cascade of breakup of the precursors of bodies in the universe, the universe appeared as a nest of bodies where an invisible world is included in microscopic bodies (e.g. atoms, molecules), which are included in minerals and rocks, which are included in satellites and planets, which are included in planetary systems, which are included in stellar systems (e.g. Solar System), which are included in galaxies (e.g. Milky Way Galaxy), which are included in galaxy clusters, and so on and so forth until the highest level of organization of the galaxy superclusters (all of the largest galaxy clusters embedded in the global system formed by the universe) is reached (Fig. 3).

Fig. 3: Inclusion levels of the systems of bodies in the universe

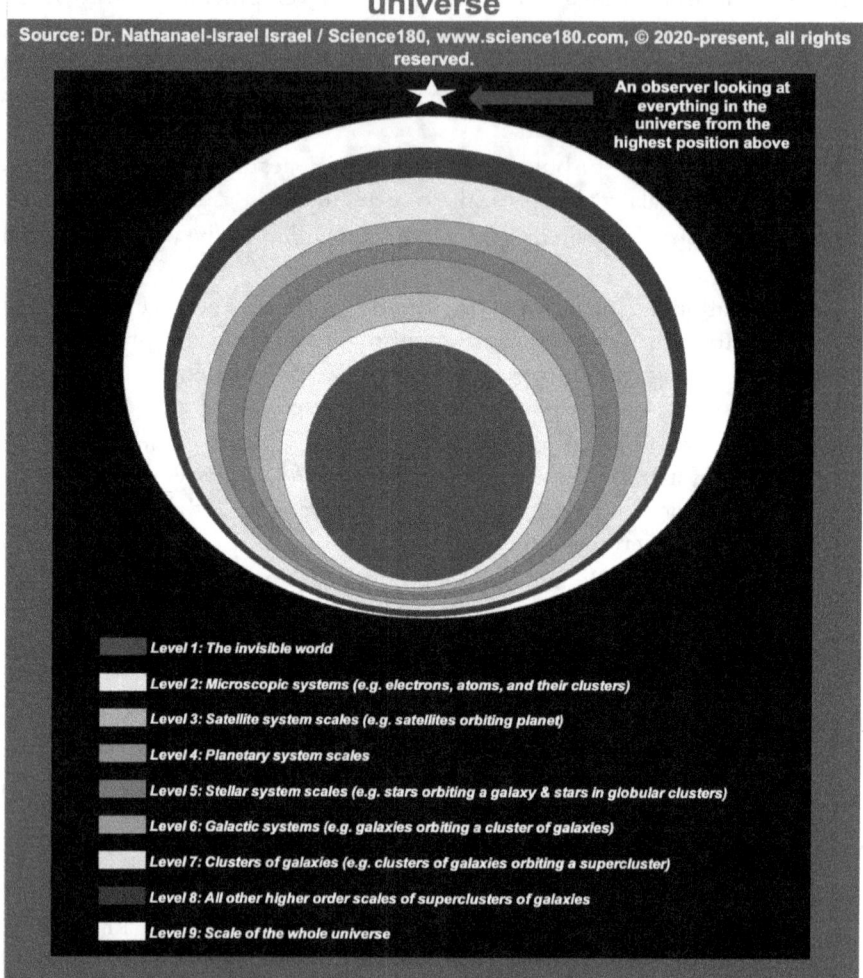

An observer looking at everything in the universe from the highest position above

Level 1: The invisible world

Level 2: Microscopic systems (e.g. electrons, atoms, and their clusters)

Level 3: Satellite system scales (e.g. satellites orbiting planet)

Level 4: Planetary system scales

Level 5: Stellar system scales (e.g. stars orbiting a galaxy & stars in globular clusters)

Level 6: Galactic systems (e.g. galaxies orbiting a cluster of galaxies)

Level 7: Clusters of galaxies (e.g. clusters of galaxies orbiting a supercluster)

Level 8: All other higher order scales of superclusters of galaxies

Level 9: Scale of the whole universe

10.2. Precursors of primary bodies and secondary bodies

As I looked at the data on the clusters of matter in the universe, I noticed that they mostly consist of a kind of "central body" (usually called the primary body) "around" which some secondary bodies are positioned. The position of the "central bodies" is roughly near what can be labeled as the "center" of the system they dominate. The primary bodies are generally "orbited" by secondary bodies in their system. For instance, electrons are believed to move in a shell around the nucleus while satellites orbit their primary planets. At their turn, planets and asteroids orbit their primary star which in the Solar System is the Sun. The Sun (if not the Solar System), in its turn, is postulated to be moving around "the" center of mass of the Milky Way Galaxy, the galaxy that the Sun belongs to. When I looked at the pictures of galaxy clusters, I felt like some galaxies may be orbiting the "center" of the cluster they belong to.

By studying the variation of the characteristics of the secondary bodies in each system and comparing them across many systems of bodies, I realized that the precursors of bodies and of systems of bodies were early split-gathered together into the precursors of up to 2 types of daughter bodies:

- the precursor of the primary body or precursor of the system of primary bodies and sometimes
- the precursor of the secondary body or precursor of the system of secondary bodies (if any)

In other words, when a mother precursor was broken up or fragmented, it yielded the precursor of a primary daughter and the precursor of a secondary daughter. While in some instances more than one primary body and more than one secondary body could have formed, in other cases no secondary body formed at all. The mathematics I did on many variables showed that the proportion of the mother precursors that landed into the primary bodies and into the secondary bodies was not by chance, but it followed specific laws that, due to space constraints, I will not detail here, but in my books *"Turbulent Origin of the Universe"* and *"From Science to Bible's Conclusions"*. Nevertheless, in the next chapter, I will address some of these laws that are crucial to understanding the code of the formation of the universe.

At the early stage of the development and differentiation of the precursors of the bodies in the universe, no defined shape had been acquired yet. As the precursor of a primary body was being molded into a primary body, it could have stayed near the "position" of its mother precursor, while the precursor of the secondary bodies could have been shifted farther away by a turbulent flow whose characteristics and development would determine the nature of the secondary daughter bodies. The destiny, the scaling of the movement, and other characteristics of the daughter bodies depended on the initial conditions and the mechanism that split-gathered their precursors and other factors such as the size, location, and energy of their mother precursors. The mother of some secondary precursors yielded many daughters, while others yielded just one. For instance, the precursor of the Solar System was split-gathered into 2 main precursors:

- the precursor of the primary body in the Solar System which was the precursor of the Sun, and
- the precursor of the bodies orbiting the Sun.

Likewise, the precursor of some atoms was split-gathered into the precursor of the nucleus and the precursor of the electrons, which would later "orbit" or surround the nucleus at specific distances from it. Similarly, the precursor of a planetary system was split-gathered into the precursor of a planet and the precursor of its satellites through various processes that people have failed to re-engineer (Fig. 4). But by the grace of God, I discovered the process and I will explain to you very soon. Praise be to God. Alleluia!

Just as some particles and celestial bodies are not systems of bodies, but just one single body, so also were their precursors. For instance, although many galaxies are found in the universe, many stars are known not to belong to any galaxy but are isolated and sometimes in clusters (e.g., globular clusters) different from galaxies. Similarly, some smaller bodies are isolated rather than associated with others in a system of bodies. In the Solar System for instance, some planets (e.g., Mercury and Venus) do not have a satellite. In other words, the precursors of Mercury and Venus were not split into a primary body orbited by a secondary body, but they were just molded into one body: a planet. The precursor of the system of secondary bodies in each planetary system is what was split-gathered into the satellites around each planet (Fig. 4). Likewise, although some asteroids are known to have their own satellites, most of them do not have a satellite. On the scale of microscopic particles, while atoms are known to have a nucleus "orbited" by at least an electron, many other types of particles found in nature are not orbited by others. For instance, electrons and other types of leptons (e.g., muons) can also be found in nature, not around a nucleus, but in free and isolated forms. In my book on the origin of chemical particles, I delved into the details.

The secondary bodies did not just start "orbiting" the primary body in their system overnight or by chance, but complex mechanisms were deployed so that, as the mother precursors were being molded into their daughter bodies, the changes that the daughters went through caused them to acquire specific movements. In general, the descendants of a mother precursor tend to cluster around where their mother has been just as children gravitate around their parent, which, in turn, orbits or gravitates also around their own parents, and so on and so forth. This explains why satellites orbit their primary planet, which, in turn, orbits a primary star, some of which, in turn, can orbit the center of their galaxy. For instance, the Moon orbits the Earth, and the Earth orbits the Sun, and the Sun orbits (or is believed to be orbiting) the Milky Way Galaxy. However, because of how they broke away from their mother precursors, some daughter precursors are unable to properly orbit their mother, yet their trajectories can still be traced back to their origin. It is as if a child can be weaned or made independent from his parents, yet his roots can be traced back to the parents. Just as some children stay in the neighborhood of their parents forever, some celestial bodies and matter also stay in the vicinity of their mothers. Some can be orbiting the remaining main body of their mother precursors, while others can have a non-orbital trajectory. Furthermore, some of the internal constituents of the daughter bodies are still moving just as the celestial bodies in the universe (e.g., stars, planets, and satellites) are not static. This explains why even inside microscopic bodies like atoms, for instance, some subatomic particles (e.g., electrons) are said to still be moving just like planets moving around their primary stars.

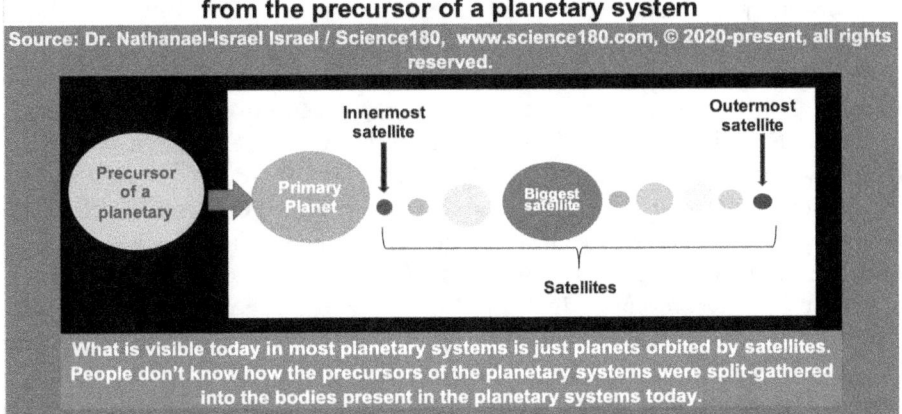

Fig. 4: Layout (as of today) of the primary planet and satellites born from the precursor of a planetary system

What is visible today in most planetary systems is just planets orbited by satellites. People don't know how the precursors of the planetary systems were split-gathered into the bodies present in the planetary systems today.

10.3. Tongue of fire in the precursors of some bodies

Although it is impossible to know the exact composition of the initial matter that God created out of nothing and which became the precursor of everything made, I think that in the process of its modification fire was one of the products. Indeed, on Friday, October 25, 2019, as I was walking home after taking my first daughter, Josephine Gabriella-Raphae, to her school, I had a strange thought that had never occurred to me before. As I was looking at the Sun that was rising in the east, it came to my understanding that, in the beginning of the universe, the precursors of all things could have been like a mega tongue of fire that split into different tongues of fire in such a way that the precursors of the stellar systems could have been a gigantic tongue of fire that was split to give rise to stars and the bodies orbiting them. For instance, the precursor of the Solar System could have been a gigantic tongue of fire that was fragmented to birth the Sun and all the bodies orbiting it. The precursors of the bodies orbiting the Sun could have also been like tongues of fire of different sizes, some of which also split to birth a primary body (e.g., planets) and bodies orbiting them (e.g. satellites). As these tongues of fire moved through their environment and took shape, some cooled (at least at their surface), yielding a solid crust with a fluid interior as a leftover or a testament to their origin. Other tongues of fire were so gigantic that the cold of their environment could not cool down their surface to yield a solid surface, hence they just took their form and remained as stars.

The constitution of these tongues of fire was different from the constitution of fire known today. For, in the beginning, as these initial tongues of fire were moving and traveling from their origin to their orbiting destination, their constitutions dynamically were changed, therefore yielding different bodies with various constitutions. As of today, when the crust of any solid body or even liquid or gas is burned, it can produce a fire as the reverse reaction of its formation, but the constitution of these fires is not the same from one type of material to the other because their chemical constituents are different. In other words, fires are like a reverse reaction of some of the events that took place during the formation of the universe. Hence, when submitted to enough energy, any material can catch fire.

As I was pondering on these ideas that were bombarding my mind as I was walking back home that day, it also came to my understanding that the biblical event that occurred during the historic Pentecost recorded in the Book of Acts, when tongues of fire were fell on the heads of the apostles who were waiting in the upper room, is a testimony of how God carved some celestial bodies and certain spiritual beings such as angels. For the Holy Spirit Himself is a supernatural fire. God Himself is portrayed as a light and a burning fire. In other words, original things or things at the beginning were made of fire which gave them energy and power. Nevertheless, God has the ability to remove that light or fire from anything He wants. Cold can extinguish fire. Cold attitudes can also extinguish good ones. For instance, after Lucifer fell and was removed from Heaven—due to his sin—, the light he initially had and which contained the power of God was withdrawn from him, hence Satan is filled with darkness which does not have (positive) power; yet Satan deceives many as if he were powerful, maybe powerful in doing evil things, which is the result of the lack of positive power. Therefore, because darkness knows that if it stands in the presence of light, it will disappear, it never likes light. Hence the Devil and his demonic agents prefer darkness and things that oppose the presence of light. Consequently, those who delight in dark or evil things bring darkness upon themselves, meaning they do things that lack the energy or fire of God. That is why God commanded his children to live in the light of God and be baptized with the fire of God, the Holy Ghost, who empowers believers with supernatural energy or ability to do mighty things. Because the word of God is light, those who read and apply it fill themselves with light and energy, which will help them to overcome evil. Even more, the Bible describes Jesus as the Light of the world, who came into the darkness to save people (even His own people), but the darkness knew Him not nor accepted Him. In contrast, the darkness ended up killing the light of the world! This is a mystery I explained elsewhere.

The presence of fire in the precursors of bodies also corroborates the revelation of Apostle Peter regarding the reverse reaction of creation which will lead to the destruction of the current universe. Indeed, in his epistles, Apostle Peter said that the Earth and its minerals will be melted in a fire, suggesting that the initial reactants that may have been used to form the Earth were fire at one point. For we know in physics that reversing a reaction produces the reactants that were initially used. Unfortunately, because so much energy and skill are needed to reverse many natural reactions or products, and because scientists do not even know how the universe was formed, they have labeled countless natural reactions irreversible. Consequently, they are not able to form, "create," replicate, or even improve many natural things, for they were made with means beyond a mere human understanding, worse when people refuse to believe in God, the Creator, they are doomed to comprehend very little.

'Science180 Academy' Success Strategy
SCIENCE180 PUBLISHING: AUTHORS WANTED

Science180 Publishing, the American publishing company that published the groundbreaking discovery about the origin of the universe, of life, and of chemicals spearheaded by Dr. Nathanael-Israel Israel, really wants to publish your book(s) regardless of your field of expertise. This is a unique opportunity for:

- established authors
- people aspiring to become authors
- people who have written a book or want to write one and need help with anything regarding publishing
- people who are not well known, inexperienced
- people whose books are viewed as nonconformist, controversial, or unconventional
- people who do not have enough resources or knowledge to navigate the publishing process
- people who are struggling to find an affordable, experienced, and high-quality publisher

Although Science180 Publishing is based in the USA, it can publish your books within your budget regardless of your geographical location. Science180 Publishing is highly interested in your document and possibly helping you publish it. Please visit Science180Publishing.com to explore how we may assist you. No matter the content of your book, as far as it is original, not promoting anything illegal, and not duplicating anyone else's idea, Science180 Publishing can help you publish it in the USA. Please contact us asap and see how we can help.

To start your journey of publishing your book with Science180 Publishing, please visit Science180Publishing.com today.

Nathanael-Israel Israel: Author of "Turbulent Origin of the Universe"

CHAPTER 11

THE 10 UNDENIABLE SCIENTIFIC FACTS THAT SET YOUR MIND TO ACCURATELY CRACK THE UNIVERSE FORMATION CODE

Which 10 scientific facts must be understood to quickly fix the tragic trend that is causing people to daringly question the Bible, while kissing secular theories that program them to believe in evolutionism, the Big Bang, and all other theories that deny God and that can never crack the code of the universe formation?

Before discovering the process of the formation of the universe, I had to scientifically study many variables, which I extensively presented in my books *"From Science to Bible's Conclusions"* and *"Turbulent Origin of the Universe."* So that the readers of this book can get a feel of these variables, I will briefly present a few below, knowing that those who want to learn more can follow up accordingly. The raw data associated with all the trends I explained in this chapter were collected by NASA (2018) and other spatial agencies in the world. As simple as the trends I will present in this chapter may sound, they represent a bedrock of codes I had to decrypt before understanding the formation of the universe. In the next chapter, I will explain the underlying cause of these trends, and then, I will delve into the scientific demonstrations of the duration of the formation of bodies in the universe.

11.1. The 99% vs. 1% rule

In my scientific book on the origin of the universe, I explained why I invented the term "9 system-additive" variables to address certain variables (for which the total value in a system can be calculated by adding the value of each of the bodies in that system) as a whole. Here, to illustrate the 99% vs. 1% rule, I will focus on just three of these nine variables: mass, volume, and kinetic energy. Mass is about the weight, volume is about size, while kinetic energy is an energy caused by the movement of a body. I demonstrated that the kinetic energy of a planetary system can be estimated by summing the kinetic energies of all the bodies in that system (e.g., the satellites and the primary planet). Similarly, the kinetic energy in the Solar System can be estimated by adding the kinetic energy of all the bodies in the Solar System.

Based on the scientific data collected by NASA, but that I had to reanalyze according to my perception of the formation of the universe, I showed that, more than 99% of the mass, volume, and kinetic energy of the bodies in the Solar System are found in the Sun, while less than 1% is found in the bodies orbiting the Sun, meaning the planetary systems and asteroids. I also showed that in the planetary systems, more than 99% of the mass, volume, and kinetic energy are found in the primary planets and less than 1% is in the satellites. In summary, more than 99% of the mass, volume, and kinetic energy in each system of bodies is in the primary body and less than 1% is in the secondary bodies. Why that trend? I will tell you in the next chapter.

Attempting to understand the laws underlying the split-gathering of the precursors of the bodies, I investigated the proportion of some variables in the primary bodies with respect to that of all the bodies orbiting them. In other words, I studied how the mass, volume, and kinetic energy (and other variables of course) of the primary bodies relate to those of their secondary bodies. As I extensively reported in *"Turbulent Origin of the Universe,"* the ratios of these variables of the primary bodies to that of the bodies orbiting them is not a constant across many systems I studied in the Solar System. When I carefully observed the distribution of the planets and satellites in the Solar System for instance, the analysis of their data based on many variables did not yield a similar distribution nor a similar law. Even when a similar trend seems to explain variation between data from one system and the other, the scaling factors of the equations differ most of the time, suggesting that the laws governing these systems are not the same, although similarities exist. Using these trends across many variables, which are the backbone of most physical laws, I showed that the physical laws are not the same everywhere in the universe. This suggests that the interactions between the primary bodies and the bodies orbiting them depend on the system. For instance, the interactions between a planet and its satellites depend on the planetary system just as the interactions between a star and the bodies orbiting it depend on the type of star. The diversity of the splitting or fragmentation of matter precursors into different clusters during the formation of the universe conferred on each cluster or body unique characteristics, which differ in scale from one cluster to another and from one environment to another. To sum it up, the laws of nature are not the same nor scaled the same across the universe. Yet, one of the major assumptions in physics and in many other scientific disciplines is that the physical laws and the distribution of bodies in the universe are the same everywhere. Therefore, basing their assumptions on similar errors, some scientists have overlooked the process by which the universe formed.

11.2. Orbital speed of celestial bodies in the Solar System

Everything in the universe is moving. Some movements are easily detectable, whereas others can be detected only with sophisticated equipment that requires extensive research. The orbital movement can be defined as the movement of an object around a primary object. In the Solar System, for instance, the movement of the planets and asteroids around the Sun is an orbital motion, also called a revolution. The orbital speed is the speed of a celestial body when considering its orbital movement. The orbital speed of the celestial bodies is not constant, but varies with their position.

I studied the mean orbital speeds of 461 bodies in the Solar System, and they range from 0.123 km/s to 47.36 km/s. The highest orbital speed was recorded with Mercury while the smallest was with Hydra, the outermost Plutonian satellite. The mean orbital speed of the planets of the Solar System ranges from 4.67 km/s to 47.36 km/s, the lowest value was recorded with Pluto (the outermost planet in the Solar System), and the highest value was with Mercury (the innermost planet in the Solar System). The mean orbital speed of the Earth is 29.78 km/s. The orbital speed of the Sun (19.4 km/s) is smaller than that of the 4 innermost planets (Mercury, Venus, Earth, and Mars), but higher than that of the 4 giant planets in the Solar System (Jupiter, Saturn, Uranus, and Neptune). The orbital speed of the asteroids and planets is negatively correlated with their semi-major axes, which is the average distance from their primary bodies (Fig. 5).

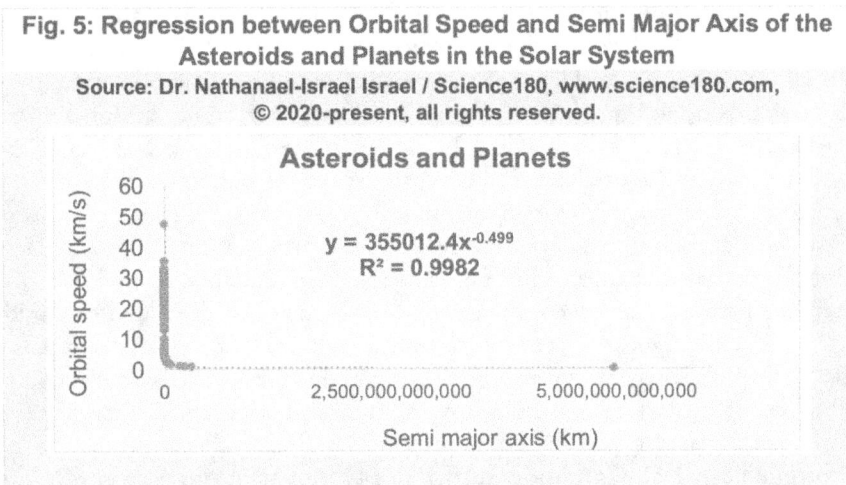

Fig. 5: Regression between Orbital Speed and Semi Major Axis of the Asteroids and Planets in the Solar System
Source: Dr. Nathanael-Israel Israel / Science180, www.science180.com,
© 2020-present, all rights reserved.

Asteroids and Planets

$$y = 355012.4x^{-0.499}$$
$$R^2 = 0.9982$$

Likewise, a strong correlation exists between the orbital speed and the semi-major axis of the satellites when I analyzed them according to their types (i.e. Martian, Jovian, Saturnian, Uranian, Neptunian, and Plutonian). In fact, the orbital speed of the satellites (Fig. 6) varies between 0.123 km/s and 31.58 km/s and depends on their types; the 4 innermost Jovian satellites have orbital speeds higher than that of the Sun.

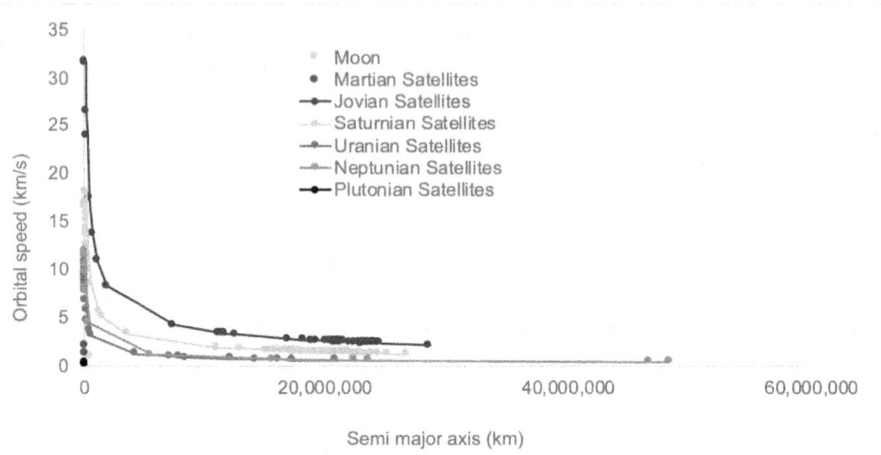

Fig. 6: Orbital speed of the satellites according to their types and semi major axis

The insight I gained from studying the scaling of the graphs I created from the data led me to reflect on what I called the law of "turbulent heredity" in living and nonliving things. I detailed this in another book. That law explains how things in nature are so interconnected that the parents of both nonliving and living things influence their children's characteristics. For instance, the lower orbital speed of some satellites is due to the fact that the kinetic energy of their precursors was too small or was positioned in a way that did not allow them to get a higher speed. In other words, the orbital speed of the bodies in the Solar System depends on the energy that their precursors had. Similarly, the speed of daughters depends on that of their mother. For instance, the vigor of seeds, no matter if they are plant seeds or animal seeds, depends on the genetic makeup of their parents. This is a kind of "heredity" law that conditions the characteristics of daughters by those of their parents. Take for instance a seed, the strength or genetics of the parents is hidden in the seed so that the daughters or children can profit from it, no matter if it is good or bad. To explain this law in another context, the heritage that children receive from their parents plays a huge role in what they will become. For instance, those who come from wealthy families often "make it" in life quickly, whereas those born into poor families usually face more difficulties before achieving "success". In other words, how fast people or things can move in life does not usually depend on the person trying to cause the movement, but on how that person himself has been launched into his life journey. For instance, no matter how much grass a mouse can eat, it will never become as big as an elephant. Similarly, no matter how much food a tortoise will eat, it can never run faster than most animals in the wild. Although human beings may use their free will to try to move themselves or other things faster, they cannot do it beyond certain limits without damaging themselves or the things they try to move. After all, everything, including human beings and their will, desire, and technologies, is limited by the constraints imposed on their environment and on the elements that are in them, including the particles constituting them. In the same manner, the movement of nonliving things is highly defined by how they were born or formed. Speed does not just depend on size. Particles or bodies born to parents full of energy and well-positioned ended up moving faster than those whose parents were less energetic or positioned at the bottom. Unless, scientists can realize this universal truth, they will continue to force some particles or bodies to move faster than they can, and in the end, these scientists can hurt themselves or the environment because of the response of these particles to the forces that are acting on them, against their nature.

11.3. Rotation period and rotation speed of celestial bodies in the Solar System

I studied the rotation periods of 224 bodies in the Solar System and they range from about 3 minutes to 243 days. The smallest rotation period was recorded on an asteroid, whereas the highest rotation period was obtained with Venus, a planet. The rotation period of the Moon (27.4 days), and the Sun (25.38 days) are much higher than that of the Earth (1 day). More than 75% of the bodies I studied in the Solar System have a rotation period shorter than 24 hours, meaning they complete a rotation in less than 24 hours (which is the time the Earth takes to complete a rotation around its rotational axis). The rotation periods of the planets in the Solar System range from 9.93 hours to 243.7 Earth days, with Jupiter having the shortest rotation period and Venus the longest. The rotation period of the 4 giant planets (Jupiter, Saturn, Neptune, and Uranus) is smaller than a day. The rotation period of the satellites varies between 2.7 hours and 79.32 days. In most planetary systems, the rotation period generally increases from the innermost satellite all the way to about the largest satellite, and afterwards, it drops. In other words, outward of the largest satellite in most planetary systems, the rotation period is very small, meaning that satellites located outward of the largest one spin faster.

Knowing the rotation period and the size of the celestial bodies, their rotation speed can be calculated. I will now briefly talk about the rotation speed of the celestial bodies. I studied the rotation speed of 322 bodies in the Solar System and found that it varies between 0.03 km/hr and 12.57 km/s, the highest value being recorded on Jupiter. The rotation speed of the Sun is slower than that of the 4 giant planets in the Solar System (Jupiter, Saturn, Neptune, and Uranus). One may wonder why the rotation speed of the Sun (1.994 km/s) is slower than that of many planets despite the huge size of the Sun. Furthermore, all the planets in the Solar System orbit or revolve around the Sun in the same direction that the Sun is rotating, which is counterclockwise when viewed from above the north pole. Why these trends? I will later explain how the direction of this rotation relates to how the precursors of the bodies in the Solar System moved or were launched.

11.4. Eccentricity of celestial bodies in the Solar System

Eccentricity is used to define the shape, circularity, and size of the orbits of celestial bodies (Fig. 7). It indicates how round or elliptical an orbit is. When an orbit is a perfect circle, its eccentricity is 0. When eccentricity is closer to 1, the orbit becomes flatter. The orbits of the bodies that have a larger eccentricity are more elongated. Here, I will focus on just one feature of the eccentricity of the satellites in the Solar System: the highest eccentricities of the satellites in most planetary systems were usually observed with the satellites located beyond the largest ones in their planetary system. In contrast, the satellites located inward of the largest ones in their planetary system usually have the least eccentricity. Why this trend? I will tell you very soon.

Fig. 7: Eccentricity of the orbit of celestial bodies

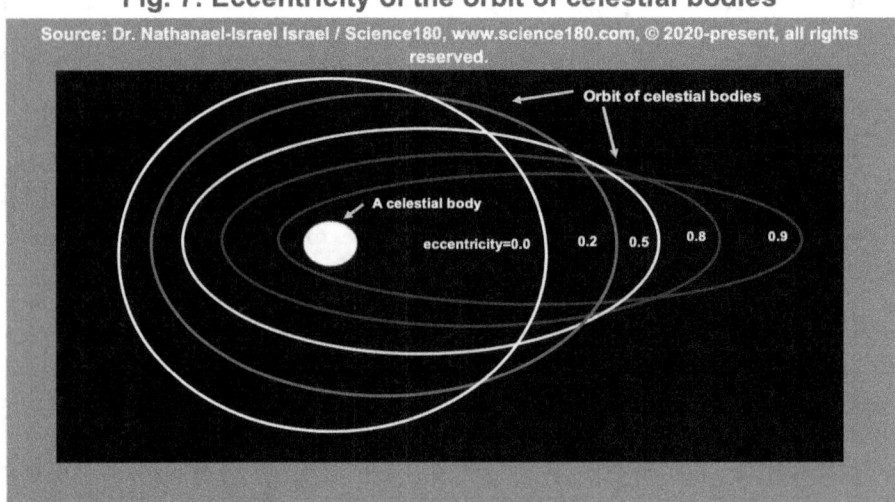

11.5. Axial tilt of celestial bodies in the Solar System

Also called obliquity of the orbit, axial tilt is a physical characteristic that defines the angle between a celestial body's rotational and orbital axes. When the axial tilt of a celestial body is less than 90°, the motion of that body is said to be prograde. In contrast, when the axial tilt is over 90°, the motion of that body is retrograde. For instance, six out of the nine planets in the Solar System have a direct rotation, meaning they rotate west to east. In contrast, three planets (Venus, Uranus, and Pluto) are retrograde because they rotate westward (from east to west). In fact, the axial tilt of the celestial bodies that I studied in the Solar System ranged from 0° to 177.36°. The lowest value was measured on a satellite, whereas the highest was measured on a planet. The least tilted planet is Mercury, whereas the most tilted planet is Venus (Fig. 8). The tilt of the Earth is 23.44°. Although all the planets in the Solar System orbit or revolve around the Sun in the same eastward direction, which is the direction of the rotation of the Sun itself, they do not rotate their rotational axis in the same direction.

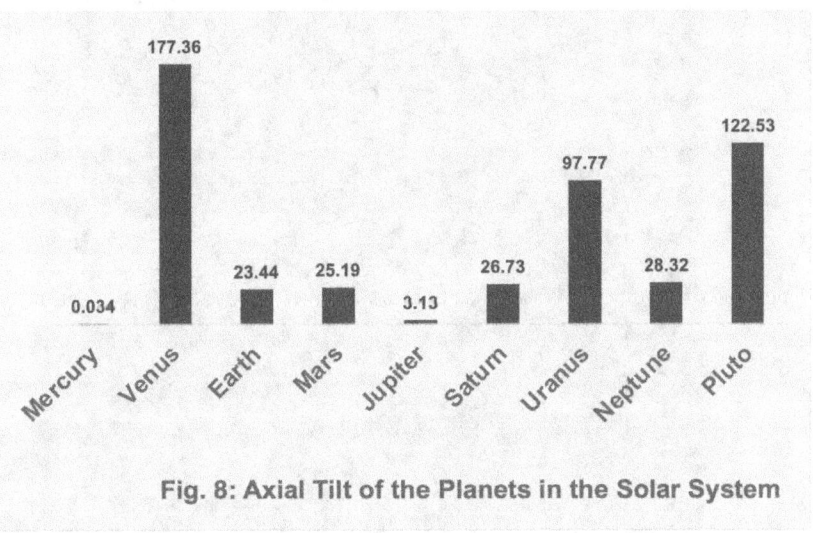

Fig. 8: Axial Tilt of the Planets in the Solar System

When ideas about the formation of the universe started flowing in my mind without me planning it, images of the axial tilt of the planets were one of the few trends that shaped some of my early perspectives on the origin of celestial bodies. In 2013, as I was looking at the data of axial tilt illustrated in the picture below, I felt like something must have pushed or pulled the planets, hence the various tilts of their rotational axes, making the motion of some bodies prograde and that of others retrograde, as if these bodies had been rolled over. Below is the sketch I drew in my notebook in 2013, when I was first inspired by the idea that the precursors of the bodies in the Solar System could have originated near the current location of the Sun. In that sketch, I mentioned that a precursor body was split, yielding other bodies positioned on different orbits. In those days, I did not really know what happened, but I was sensing that what happened was accompanied by a huge wind. Hence, I mentioned the wind, called "vent" in French, in my note below (Fig. 9).

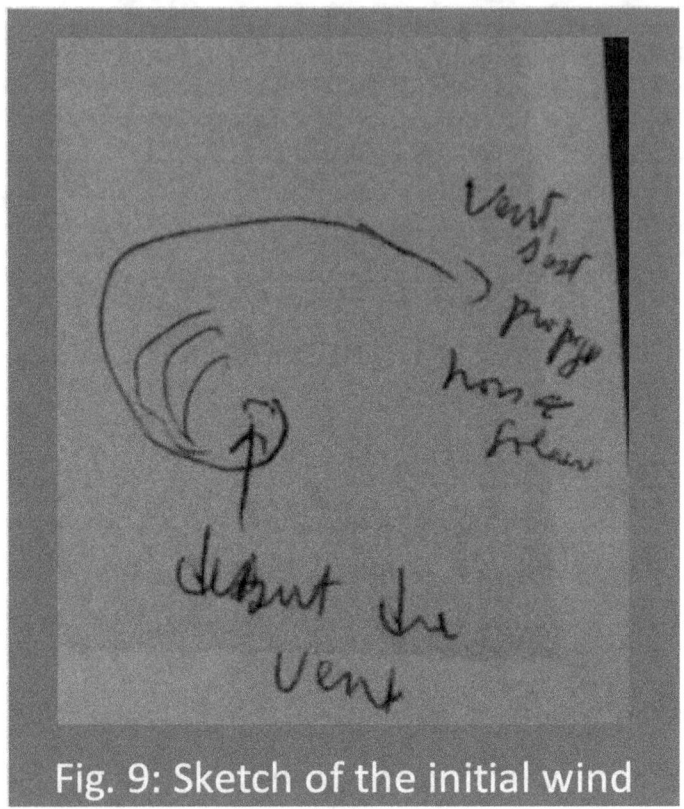

Fig. 9: Sketch of the initial wind

Now, I will talk about an insight I got from a biblical verse on tilt, which also shaped my early perspective on axial tilt. Indeed, as early as December 18, 2013, I was intrigued by some information the Bible gave on the tilt of the Earth and/or the tilt of other celestial bodies. For instance, according to Job 38:37, *"God tilted (tapped over) the water jars of heaven so that dust (gas) can become a mass of mud and its clods stick together"*. This statement implies that, during the creation of the universe, the tilting of the precursors of some celestial bodies including the Earth, which originally was a mass of fluid (e.g. water), caused that mass to amass and stay together. I deduced that fluids in the form of water and gas were present at an early stage of the formation of the Earth and other planets or celestial bodies. I also deduced that the tilting of the precursors of the bodies may have also caused them to stick together or be converted or compressed into different types of gas, liquids, and solids. Although I did not know the phenomena that led to the tilting or inclination of the precursors of the bodies in the universe, I always knew that something inclined them and their movement went through a process that led some to become gas, others to become liquid and others to become solid in some cases. The insight I got from the above biblical verse shaped my early perspective of how the tilting of the

102
Nathanael-Israel Israel: Known as the #1 International Authority That Truly Unlocked the
Secrets of the Turbulence that Shaped the Universe

precursor of the celestial bodies could have affected their formation.

11.6. Orbital inclination of celestial bodies in the Solar System

Celestial bodies in the Solar System moved on different orbits generally defined by a plane. Orbital inclination is a parameter or variable used to define the tilt of the orbit. Called the elliptic plane or the elliptic, the plane of the orbit of the Earth is the reference plane used to define the orbital inclination of the planets and asteroids in the Solar System. Hence, the orbital inclination of Earth is generally considered to be zero. The orbital plane of the satellites is usually measured with respect to the equator of their primary planet. The orbital inclination is also used to define the motion of celestial bodies. Indeed, when the orbital inclination is between 0° and 90°, the motion of that body is prograde, meaning that the body and its primary body are rotating in the same direction. In contrast, when the orbital inclination of a body is between 90° and 180°, the motion of that body is retrograde, giving the perception that the body is rotating in a retrograde or reverse direction, which is opposite of its primary body (Fig. 10).

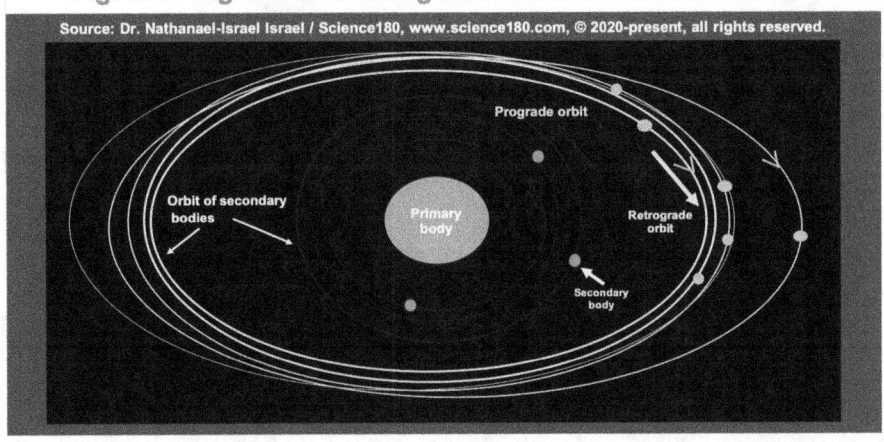

Fig. 10: Prograde and retrograde orbits of celestial bodies

The orbital inclination of the planets is generally small, whereas that of the comets and the other outermost celestial bodies is high. The most important trend I want you to keep in mind here is the orbital inclination of the satellites. Indeed, in each planetary system, the orbital inclinations of most satellites outward of the largest satellite in their system are the highest, while the orbital inclinations of the satellites inward of the largest satellite (in their system) are the smallest. In other words, in each planetary system, the orbit of the satellites inward of the largest one is almost flat, while that of most satellites located outward of the largest satellite is highly tilted. Nevertheless, a few retrograde satellites are found between prograde satellites. The high value of the orbital inclination of the outermost satellites suggests that, at one point, something acted on the precursors of these satellites to tilt the plane of their orbit.

An unanswered question in astronomy is the cause of the variation in the inclination of the celestial bodies. Before my discoveries about the formation of the universe, no profound work was done to explain why some bodies are more inclined than others, or why some bodies are inclined at all. So far, all of the theories elaborated before my findings just speculated that the inclinations of the celestial bodies were caused by random collisions during the formation of the universe. But very soon, I will tell you that this trend was not caused by collusion but by another factor that was key to unlocking the code of the universe's formation.

Furthermore, to finish this segment, I would like to say a few words about the inclination to the Sun's Equator. Indeed, in addition to orbital inclination, the inclination to the Sun's Equator is another parameter used to characterize the plane of the orbit of the planets. Finally, although it is commonly said that the planets in the Solar System are on the same plane as the Sun's equator, it is important to note that this statement is incorrect. For instance, the orbital planes of the planets in the Solar System are inclined differently with respect to the Sun's equator. The inclination to the Sun's equator of the planets varies from 3.38° to 11.88°, the lowest value being recorded with Mercury, while the highest value is with Pluto. Earth's orbital plane is tilted 7.25° with respect to the Sun's equator. However, because the Sun wobbles during its motion, the inclination to the Sun's equator slightly varies with time. Because the orbit of the planets in the Solar System is NOT in the same plane as the Sun's equator, all the theories that claim that the planets and the Sun are in the same plane are simply false.

11.7. Density of celestial bodies in the Solar System

Expressing the amount of weight found in a certain amount of space, meaning the ratio of mass and volume (e.g., kg per meter cube), density is used to measure the degree of compactness of things including celestial bodies. The density of the celestial bodies I studied ranges from 250 kg/m^3 to 6730 kg/m^3. Many of the smallest densities were recorded on comets and Saturnian satellites (Saturn itself being among the least dense bodies in the Solar System). In general, the densest bodies were recorded with the main belt asteroids and planets. The densest planet is Earth, whereas the least dense planet is Saturn (Fig. 11). The terrestrial planets are denser than the Sun and the giant planets (Fig. 11). Among the planets, the lowest densities are found in the giant planets. In my book *"Turbulent Origin of the Universe,"* I proved that Saturn's low density is also connected to the formation of the Saturnian rings, the largest rings recorded on any planet in the Solar System. Finally, the density of the Sun (1408 kg/m³) is smaller than that of 58% of the bodies I studied in the Solar System. In other words, the precursor of the Sun was not as compressed as the precursors of the other bodies in the Solar System. The key point I want to underline in this segment is that the largest bodies are not the densest ones, but they are some of the least dense. Later, I will explain this trend.

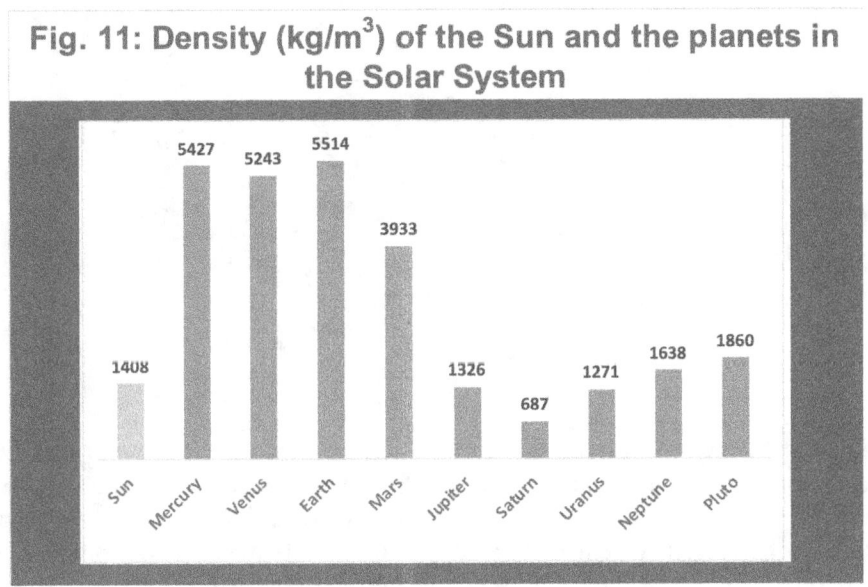

Fig. 11: Density (kg/m³) of the Sun and the planets in the Solar System

11.8. Semi-major axis of celestial bodies

The semi-major axis of a planet in the Solar System is its mean distance from the Sun from the center (of the planet) to center (of the Sun). For the satellites, the semi-major axis is the mean distance from their primary planet (from the center of the satellite to the center of the planet). The semi-major axis measures the size of an orbit and is the longest radius of the ellipse of the orbit of the celestial bodies. Also used to measure the distance between bodies, the Astronomical Unit (AU) is the mean distance from the Sun to the Earth: 149,597,900 km. The semi-major axes of the planets in the Solar System range between 0.387 AU and 39.482 AU (Fig. 12), with the smallest value being recorded with Mercury and the highest value found with Pluto. Varying between 9,378 km and 48.4 million km, the semi-major axis of the satellites is much smaller than that of their planets. Celestial bodies are located at specific distances from one another, and particularly from their primary bodies.

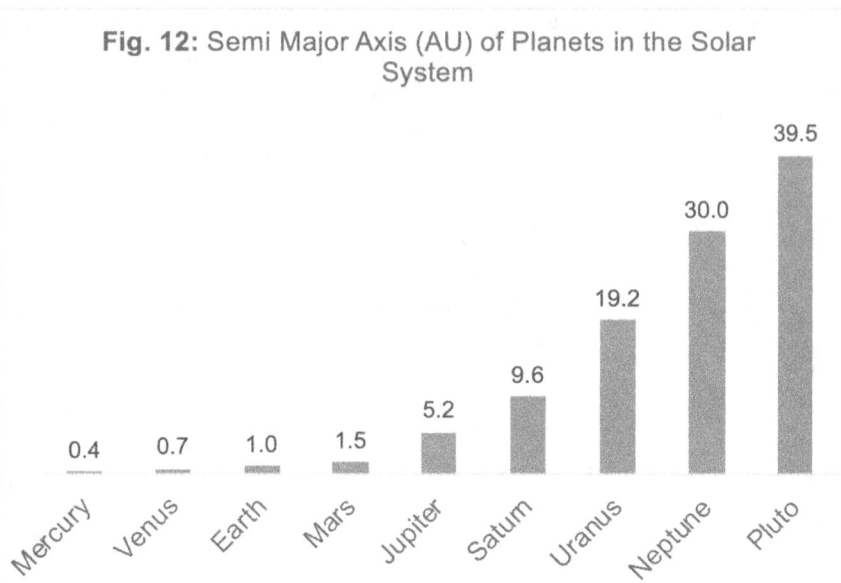

Fig. 12: Semi Major Axis (AU) of Planets in the Solar System

Now, let me say a few words about the increment of the semi-major axis of the bodies in the Solar System. Indeed, the semi-major increment between two points A and B is the difference between the semi-major axis of B and the semi-major axis of A. In other words, the semi-major axis increment between two bodies can be calculated by subtracting the semi-major axis of the body that is upstream from that of the body immediately downstream. In the Solar System, the semi-major axis increment ranges from zero to millions of kilometers, meaning that some bodies shared almost the same orbit, while some are millions of kilometers farther from their nearest neighbors.

Furthermore, scientific evidence suggests that the universe has been expanding, meaning that the distance between some bodies has been increasing. The Bible also talks about the expansion of the universe, and attributes it to the creative work of God. For instance, talking about the stretching of the universe, God told Prophet Isaiah in Isaiah 45:12: "*I* [God] *made the Earth and created man on it. My hands stretched out the heavens, and I commanded all their hosts.*" Because of this expansion, the distance between some constituents in the universe may be increasing. Therefore, the distance between some celestial bodies as of today is the combination of (1) the distance between them at the end of their formation and (2) the distance caused by the expansion that occurred at their level afterwards.

11.9. Escape velocity

Because of gravity, anything on Earth is bound to Earth unless a sufficient force is used to propel it out of the atmosphere. Similarly, most celestial bodies that are big enough tend to hold anything that is on them or in their atmosphere... The gravitational pull is defined as the force that causes some celestial bodies to attract other objects toward them. To pull something out of a celestial body's zone where the gravity rules, an energy must be applied to allow that thing to move out. According to NASA, the escape velocity is the *"initial or minimum velocity that is required at the surface (or at the 1 bar pressure level for the gas giants) to escape a body's gravitational pull, while ignoring atmospheric drag"*. Once the escape velocity is applied to any object, that object is supposed to escape the gravitational pull of the celestial body where it was. For instance, if you throw something in the air, it will likely fall back on Earth. However, if the escape velocity is applied to that thing, it will escape the Earth, meaning that it will leave the Earth and go into space and never return to Earth because of Earth's gravity. According to NASA (2018), the escape velocity of the Sun is 617.6 km/s, while that of the Earth is 11.18 km/s. The insight that scientists gained into the escape velocity has helped them to properly launch some space engines. How about the role of the escape velocity on the way the precursors of celestial bodies were launched? Later in this book, I will show how the escape velocity helped me to calculate the duration of the formation of the celestial bodies in the universe. More details about escape velocity can be found in my book *"Turbulent Origin of the Universe."*

11.10. Radius of celestial bodies in the Solar System

I studied the radius of 440 bodies in the Solar System. The Sun is the biggest object in the Solar System, and its radius is 696,000 km. Planets are the biggest objects orbiting the Sun, and their radius ranges between 1,187 km and 71,492 km (Fig. 13), with Jupiter being the largest planet in the Solar System. The radius of Jupiter is 10.27% that of the Sun and 11.2 times that of the Earth, meaning that the radius of the Earth (6378.14 km) is about less than one-hundredth that of the Sun.

Science180: Efficient, Trustworthy, and Cost-Effective Company to Add to Your Strategic Journey Toward Your Best Tomorrow

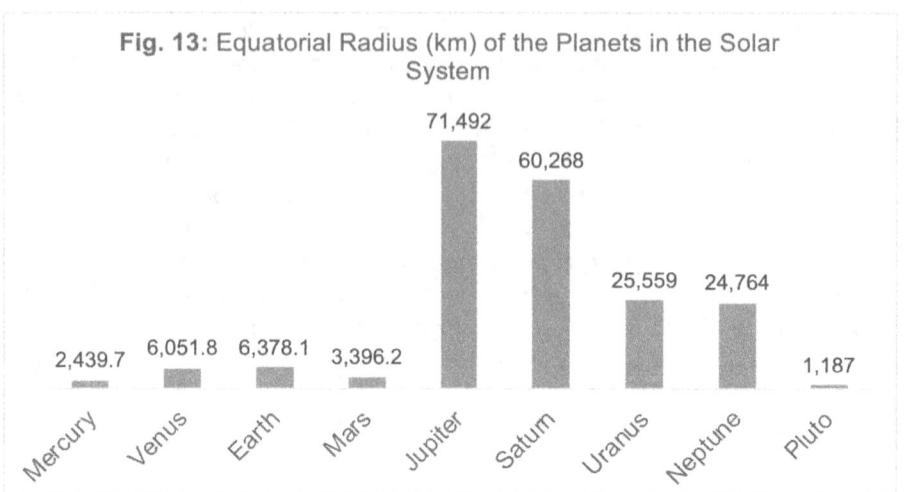

Fig. 13: Equatorial Radius (km) of the Planets in the Solar System

The radius of the satellites in the Solar System varies between 300 meters and 2634.1 km. The largest satellites are neither the innermost nor the outermost satellites but are located some distance in between. In *"Turbulent Origin of the Universe"*, I proved that the biggest satellite in each planetary system belongs to a region I called the most turbulent zone (e.g., turbulence Zone 3), while most of the smaller satellites are found at the edge of their planetary system. Fig. 14 shows the radius of the satellites.

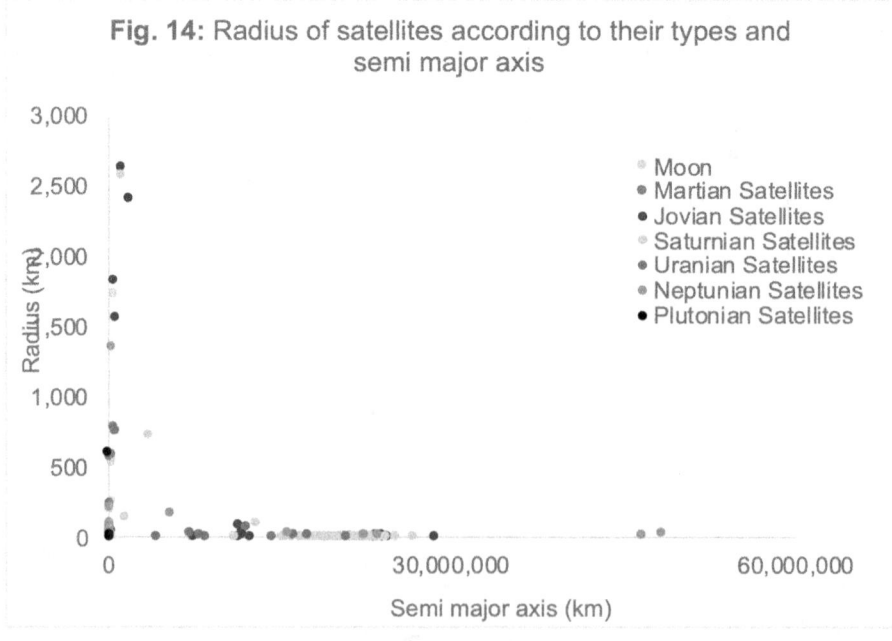

Fig. 14: Radius of satellites according to their types and semi major axis

CHAPTER 11: TEN UNDENIABLE SCIENTIFIC FACTS TO CRACK THE CODE OF THE UNIVERSE FORMATION

As I am discussing the radii of satellites, I would like to mention that the 4 largest Jovian satellites (also known as the Galilean moons) were the first objects found to orbit a body other than the Earth or the Sun. The German astronomer Simon Marius was credited for independently discovering these 4 major Jovian moons, and the names used for those moons today are those given by Simon, not those given by Galileo. The discovery of these Jovian satellites dealt a blow to the then-accepted Ptolemaic world system, a geometric theory that claimed that everything in the universe orbits the Earth. At that time, the Copernican system, which claimed that everything in the universe orbits the Sun was also prevalent. To compare the Copernican and Ptolemaic systems, Galileo wrote a book in 1632, which is said to defy or attack the (wrong) astronomical belief of the then-pope, Pope Urban VII. The Pope and the Jesuits felt alienated, and consequently, Galileo was sentenced to house arrest for life for what was called heresy, therefore silencing Galileo (known as the father of observational astronomy, the father of scientific method, and the father of modern physics) until his death 10 years later. Galileo's book was condemned and placed on the list of forbidden books. By the way, do I need to say that, even until today, many Jewish books holding scriptures (e.g. the Books of Enoch, which I came to realize contain many secrets about the formation of the universe) were also forbidden and rejected by most Christians, Jews, and those who don't even believe in Christianity or Judaism? Knowing and learning from all of those things and others, I proceeded with the publication of my works strategically and very carefully. In the next chapter, I will explain how the trends I addressed in this chapter can be tied together.

Science180: Efficient, Trustworthy, and Cost-Effective Company to Add to Your Strategic Journey Toward Your Best Tomorrow

'Science180 Academy' Success Strategy:
SCIENCE180 SERVICES AND PRODUCTS YOU WILL LOVE

Because you are reading this book, you are probably very interested in answering your questions about the origin of the universe, of life, and of chemicals. Imagine you want to be trained by Dr. Nathanael-Israel Israel and his team so you can benefit from their outstanding expertise to empower yourself or your team. Or you want him to give a keynote speech, a seminar, or any other kind of talk or conference at your organization. Or you want him to mentor you or some people or teams at your organization. Maybe you have critical origin-related questions that you need his help to accurately answer. You want a true expert to talk with you about the customized program or game plan that fits your needs. You want him to tailor his advice, expert feedback, and proven shortcuts to the stage of life you are in and help you get to where you want to be in your desire to properly understand the origin of the universe, life, and chemicals and harness the benefits that come with it. Perhaps you don't know how to properly complete these important tasks according to your specific needs or your organization's needs and demands. That is what Science180 Academy is all about. Visit Science180.com/services for more details about how to benefit from the services that Science180 provides.

Maybe you are a leader who wants to hire Dr. Nathanael-Israel Israel and his team to train some departments at your organization. Or you want to refer them to other companies like a good dish passed around the dinner table, and you want to explore how Nathanael-Israel Israel can pay you something for that referral. Maybe you attended Nathanael-Israel Israel's speaking program, for which, without going into details, he accurately raised your awareness about how the universe, life, and chemicals were formed. Or maybe you attended his training, in which he detailed and showed you how he decoded the scientific data using various tools and certain thinking strategies that helped him and which transferred some skills to you; and now, you are interested in a long term one-on-one consulting, or mentoring program with him, so that, he delves into more details about how to use proven techniques to decode the universe (strategies for data collection, data analysis, data presentation, writing, and even tips for future research) and change your behavior on a long term basis. If you relate to any of the points mentioned above, Science180 Academy is the right fit for you!

Other customizable services that Science180 provides include Assessments, books, and other products, publishing your book, Conferences, Consulting, Executive mastermind groups, Face-to-face visits, Master classes, Online courses, Podcasting, Seminars, Speaking engagements, Survey and research tools, Training, Video programs.

CHAPTER 11: TEN UNDENIABLE SCIENTIFIC FACTS TO CRACK THE CODE OF THE UNIVERSE FORMATION

Science180 is the only company that has scientifically, completely cut the rational legs upon which all God deniers, all modern freethinkers, evolutionists, and Big Bang proponents had wished to stand to continue rejecting God. Science180 is the only company that left all wrong creationist proponents, evolutionists, Big Bang advocates, and all other anti-creationists that do not literally believe the Biblical creation in an echoed silence, head-scratching and asking: "Who took our secular and religious spotlight?". The vision of Science180 is to be the indisputable, accurate voice that engages and enlightens the global effort that unconventionally promotes the perfect agreement between science and the Biblical account of Creation in a truthful way that is highly helpful to and respected by both scientists and nonscientists, believers and all freethinkers (e.g., atheists, unbelievers, rationalists, and skeptics), adults as well as children, as a fundamental, beneficial, and free good worthy to guide them in making holistic decisions that will serve them today and forever.

Here are other reasons why you should choose to work with or hire Nathanael-Israel Israel and the team at Science180:

- Accurately understand universe-creation. Be happy forever!
- Biblical Genesis inerrancy guaranteed
- Bringing the Judeo-Christian believers together through the power of the accurate decoding and understanding of the Biblical account of creation
- Discover the key variables needed to decode the universe's creation
- Easily understand complex universe-creation equations in minutes
- Enjoy scientifically verifiable, biblically based universe-origin, life-origin, and chemicals-origin models
- Find the creation of the universe for a reasonable price
- First stop for your universe-creation needs
- Impressively accurate and impossibly easy-to-understand universe-creation theory
- Irrefutable scientific demonstration of creation
- Mover of the needle on the science vs. creationism debate
- Nonconformist, rule-breaker, and accurate demonstrator of the universe's creation
- Reuniting science and the Biblical account of creation

- Science and Bible reconciliation made possible
- The all-in-one proven & uncomplicated universe-creation formula
- The formula at the intersection of science and the Bible
- The most accurate, reliable, safest, and best explanation of the universe's creation ever
- The most trusted name in cosmology
- The new physics that will revolutionize science forever
- The place where the universe's creation gets decoded accurately
- The science that reunites your reason and faith
- The undeniable scientific challenge to all metaphorical, figurative, loose, liberal, or vague explanations of the Biblical creation
- The unquestionable scientific challenge to Big Bang & Evolutionism
- Trailblazer of the reconciliation between science and the Biblical account of Creation
- Undeniable reconciliation of science and the Biblical account of creation
- Understand the creation of the universe. Increase your glory and peace of mind
- Universe-creation formula accurately made easy
- Universe-creation theory that helps you fight wasteful programs
- Universe-decoding accuracy guaranteed
- Where universe-creation is accurately decoded, full stop

CHAPTER 12

CAN WE SCIENTIFICALLY DECODE THE UNIVERSE'S FORMATION WITHOUT USING A KEY IN THE BEHAVIOR OF FLUID LAYERS LIKE THE WATERS MENTIONED IN THE GENESIS ACCOUNT?

After discovering the trends I mentioned in the previous chapter, I wondered which story could tie them together. It took me more than 5 years before I got the answer that I summarized below. I discovered that, during the formation of the bodies in the Solar System (and in other parts of the universe), fluids existed and were organized as layers, which interacted with one another. The footprint of the processes that were involved in the molding of the celestial bodies was left in their characteristics collected by scientists throughout the ages. Here, I will not go over all the details, but I will just pick a few things that can make more sense to you, hoping that those who want the long proofs can consult my book *"Turbulent Origin of the Universe."* In that book, I showed that the precursor of the satellites escaped the precursor of their primary body at about the escape velocity of the primary planet. Likewise, the precursor of the bodies orbiting the Sun escaped the precursor of the Sun at about the escape velocity of the Sun.

I demonstrated that the precursors of the celestial bodies did not split-gather by chance, but according to specific laws. For instance, when the precursors of systems of bodies were split-gathering, more than 99% of the mass, energy, and volume went into the precursor of the primary body, while the remainder went into the precursor of the secondary bodies. Hence, more than 99% of the mass, energy, and volume of the bodies in the Solar System are found in the Sun and less than 1% in the bodies orbiting the Sun. The trend also explains why more than 99% of the mass, energy, and volume of the bodies in the planetary system are found in the primary planet, while less than 1% is in the satellites. In other words, based on the scientific data collected by NASA, but that I had to reanalyze according to my perception of the formation of the universe, I showed that, during the split-gathering of the precursor of the Solar System, more than 99% of its mass, volume, and kinetic energy went into the precursor of the Sun and less than 1% went into the precursors of the bodies orbiting the Sun, meaning the precursor of the planetary systems and asteroids. After going through some turbulent changes, the precursor of the Sun "finalized" its formation and became the Sun, which as of today has more than 99% of the mass, volume, and kinetic energy of the bodies in the Solar System. Then, during the split-gathering of the precursors of the planetary systems, more than 99% of each of the mass, volume, and kinetic energy went into the precursors of the primary planets and less than 1% went into the precursors of their satellites. In summary, the general rule of the split-gathering together of the precursors of the systems of bodies is such that more than 99% of the mass, volume, and kinetic energy were pushed into the precursor of the primary body and less than 1% was pushed into the precursor of the secondary bodies.

When the precursors of the celestial bodies were splitting from one another, fluid ligaments were formed. In each planetary system for instance, as the fluids of the precursor of the satellites were leaving the precursor of the primary planet, a neck connecting these two fluid bodies was formed. Once all the fluids of the precursor to the satellites escaped from the precursor to the primary planet, the fluid neck was broken. Likewise, a neck of fluids was formed when the precursor of the bodies orbiting the Sun was escaping the precursor of the Sun. After this ligament broke away from the precursor of the Sun, the precursor of the bodies orbiting the Sun continued its journey and began branching its fluid layers according to their positions.

I showed that the trend of the orbital speed, inclination, and eccentricity of the celestial bodies can be explained by the interactions between the fluid layers of their precursors. Indeed, scientific studies showed that, in a stack of moving fluid layers, the speed of the layers decreases from the top layer to the bottom layer—for, as the fluid layers slide over one another, each layer moves faster than the one beneath it— and the squashing of the bottom layers by the top layers stresses, stretches and spins up the bottom layers and causes them to form smaller bodies (Malvern, 2016, George, 2013, Petitjeans and Bottausci, 2020). This implies that during the split-gathering of the bodies' precursors, some fluid layers were stacked one on top of the other and slid over one another, with each layer moving faster than the one beneath it. This means that, in each system, the uppermost fluid layers had the maximum speed, while the bottom layers had the lowest speed.

Stress is the amount of force applied per area of an object. In the case of the celestial bodies, in addition to the force that propelled or pushed them away from their mothers, some fluid layers were subjected to the weight of the fluids on top of them. And from the top layer to the bottom layer, this stress increased, and caused the bottom layers to be under more pressure. Imagine yourself with a container of water on your head. The heavier the container on your head, the more pain you would feel in your neck and head, and you could not bear that weight beyond a certain capacity, or else you would be crushed. If you could be asked to move while the container is on your head, you would notice that your speed would be inversely proportional to the weight on your head. In other words, the heavier the weight on your head, the slower you would move, and the lighter the weight, the faster you could move. Likewise, the bottom fluid layers of the precursors of celestial bodies bore or felt the weight of the fluids above them. The more the stress, the slower the bodies moved. Hence in the end, because the fluid layers were later collected into celestial bodies, the speed of the primary planets is defined by their position with respect to the primary star in their stellar system, while the speed of the satellites is defined by their position with respect to their primary planet. For example, because the top fluids of the precursor of the bodies orbiting the Sun were used to form the innermost bodies in the Solar System, the orbital speed of the innermost bodies in the Solar System is higher than that of the outermost bodies, which were born from the fluid layers located at the bottom of the precursor of bodies orbiting the Sun. For similar reasons, the innermost satellites move faster than the outermost satellites and the speed of satellites decreases from the innermost to the outermost satellite.

Furthermore, as the fluid layers of the precursors were being molded into celestial bodies and positioned into their orbits, their speed was converted into orbital speed and rotational speed. Because the speed of the fluid layers of the precursor of the celestial bodies decreased from the top layers to the bottom ones, the orbital speed of the daughter bodies also follows the same trend. As the fluid layers were being split, they were wrapped around or wound up. The speed with which they were wound up decreased from the top layers to a point where their ability to rotate was very small, and consequently, the fluids around that position were stuck, leading to the formation of bigger bodies around that position. The largest satellites were formed at the place where more fluids were stuck or trapped. When the ligament of the precursors of bodies broke up, the precursors of bodies that were downstream of the breaking point could have been pushed downstream, while those upstream of the breaking point may have moved upstream. In the process, the precursors of some bodies were flipped over or overturned. The stress or strain in the fluids also affected the movement and orientation of the bodies. All these movements affected the orbital inclination of the bodies. As the fluid layers were flowing, structures formed inside of them (e.g., vortices) were also flipped, and some fluids mixed with others and rearranged. Larger ones could have swallowed some small vortices, meaning that some small fluid layers and structures were incorporated into bigger ones. The fluid layers of the precursors of bodies did not always move smoothly passing adjacent layers without mixing to some extent. Some larger bodies were born by larger fluid layers, while smaller bodies descended from thinner layers. The interactions between the unstable fluid layers (moving at different speeds) developed into turbulence. The movements of the fluids amplified the turbulence, and the balance of these turbulent movements of the bodies (e.g., vortices) as a whole contributed to birthing the rotation of their daughter bodies.

For in turbulent fluids, vortices are always formed and they usually rotate. Hence, in the end, the bodies carrying these vortices also rotate. In other words, the rotation of celestial bodies can be traced back to the turbulence of the fluids of their precursors.

The presence of fluid layers in the precursors could explain the strata found in the crust and atmosphere of some daughter bodies today. After the formation of the universe, some depositions (e.g., sedimentation) also later affected the stratification of the upper layers of some celestial bodies. For instance, some strata of rocks or minerals in the Earth's upper crust could have formed after the deposition of materials unrelated to the fluid layers of the precursors. Likewise, because of the fluid layers in the precursors, celestial bodies and even particles are surrounded by layers of fluids or layers of particles which must be crossed before reaching the surface of the bodies. That is why the Earth, for instance, is surrounded by layers of gas, including the ozone layer in the atmosphere.

The direction of flow in the fluid layers of the precursor bodies orbiting the Sun can explain the counterclockwise motion of most celestial bodies in the Solar System as seen from the north pole. In general, the fluid layers of the precursors flowed in the direction of the force that pushed them away from the initial position of their mother layer. For instance, most planets and asteroids in the Solar System orbit the Sun in the same direction, just as many satellites orbit their primary planets in the same direction. How the precursors were separated caused some of the daughter bodies to flip over; hence, the motion of some bodies is prograde, and that of others is retrograde. Considering the direction of the orbital movement in the Solar System, I showed that the dominant direction into which the initial instability could have launched the movement of the precursors of the bodies in the Solar System could have favored a counterclockwise direction, when viewed from the Northern Hemisphere.

Although the nature of the matter in the precursors of bodies in the universe is different from that of the matter today, the instability of the turbulent prima materia affected the topology of the fluids in the layers of the precursors of bodies. The topological changes of the fluid of the precursors contributed to the formation of diverse types of matter and systems of bodies in the universe. For instance, the addition or the pouring of cream into a cup of coffee destabilizes the latter, and the intensity of the instability affects the shape of the fluid "filaments" that are formed. The interfaces between the fluid layers of precursor bodies could have given rise to highly unstable structures that could have grown as they mixed and interacted with others. The internal and external changes of the precursors of bodies could have continued until they reached a point that could no longer have allowed major changes. As the clusters of bodies of different scales were interacting with one another, their degree of freedom could have been decreasing until those clusters could no longer rearrange much, therefore locking themselves into a sort of "equilibrium" stage, which also defined some of their "final" characteristics. The scale of the topological changes in the fluids of the precursors defined the scale or size of the bodies or systems of bodies in the universe. Without the topological changes, things in the universe could have been different and less diverse. If the intensity of the instability was different, a completely different world would have been formed. The structural reorganization of matters in an instable fluid depends also on the duration of the instability or the time it would take for the degree of freedom of the fluid to be significantly reduced. For instance, a certain volume of coffee mixed with a certain volume of cream can yield different structural mixtures depending on how vigorously the mixture is stirred and/or shaken.

While at one point, the bulk of fluid layers in the precursors moved together as if the layers were bound together by a yoke, at another moment during their journey, the fluid layers took different paths and separated from one another. By observing images of spiral galaxies (you can see them online), I felt that their outer stars were not wrapped as tightly, or pulled or pushed toward the center of these galaxies, as some inner stars seem to be, suggesting that the force that could have wrapped them around might have been limited. As some vortices or eddies were being "rolled" to form circular or spherical bodies, they were squeezed. The energy in the precursors may have also affected the squeezing intensity. As the fluid layers were being wrapped or curled, they could have squeezed their daughter's bodies. In the giant planets (e.g., Jupiter and Saturn) or in stars (e.g., the Sun), whose outer surfaces are not solidified, vortical structures carrying the footprint of the turbulence that occurred during their formation still exist. In the case of terrestrial bodies (those with a solid crust) for instance, the vorticial structures could have been solidified into rocks and then piled up to yield the crust.

As some precursors were being molded, their constitutive particles and the interactions between them were changing until they were locked into systems that defined their motion in harmony with their environment. One of the factors that contributed to reducing the interactions between particles could have been temperature. For instance, when the temperature was below certain values, even a very hot liquid could freeze at least on its surface, therefore limiting the interactions between some of its constituents. In contrast, when a precursor body is very big and extremely hot, its internal energy and heat could overcome the cold of its surrounding environment in a way that it cannot freeze or solidify. Instead, such huge precursors can stay very hot and maintain some of their initial heat and configuration of their constituents. The surface of some precursors could never be solid, but could emit light like stars. In other words, some stars are like a remnant of their precursor, which was too big to allow the cold in their environment to freeze them or force them to become a solid. Some stars can die if for instance their energy can no longer be sufficient to overcome the cold of their environment and/or if the constituents of their surface can undergo some modifications that can solidify them. Depending on their size, spatial localization, speed, and characteristics of their constituents, the same precursor can lead to the formation of different final daughter bodies. For instance, a liquid can be converted into ice by lowering its temperature below a certain critical value. However, depending on the duration of the temperature drop, the same water can form various types of ice with different structures or states of matter. Similarly, the outcome of some precursors was highly affected by the duration of the process they went through. No matter the explanation I gave above, it is important to retain that God is the one who gathered the Earth (Isaiah 10:14) and certainly all bodies during the formation of the universe.

As the precursors of bodies were being assembled and their movement set, they were dynamically led into their orbit, where they will spend the rest of their "lifespan." This implies that, when the bodies were being formed, their position was changing in space and at the end of their formation, the position of their orbital movement inside their system was "locked down." The initial push or launch of the precursors of matter was partially responsible for the so-called "expansion of the universe." Even after their formation, celestial bodies still have a residual tendency to move away from the position of their mother precursors. In *"Turbulent Origin of the Universe,"* I detailed the processes involved in the positioning of the bodies.

By the end of the universe's formation, energy was condensed into matter and matter systems. Part of the energy in the turbulent prima materia was used to form matter, while others were stored in different components and bonds between those matters. Energy was also used to tilt, incline, or flip over the orbital plane and rotational axis of some bodies. Chemical energy has been stored in atomic bonds and can be released and/or absorbed during chemical reactions. Energy is also stored in the nucleus of atoms and can be released during nuclear reactions. Mechanical energy was stored in matter as their precursors were wrapped and shaped. How a matter was wrapped could have also determined the amount of energy it contains. This can explain why some of the biggest chemical elements like uranium and plutonium, are some of the most energetic ones. For the sake of time and by necessity of not loading the nonscientists with too many scientific things, I decided not to delve any longer into the energy here. For more information about the forms of energy in the universe and how they point to the process of the formation of the universe, please visit *"Turbulent Origin of the Universe."*

CHAPTER 13

ONE SIMPLE SCIENTIFIC RECIPE THAT WILL MAKE ANYONE PAY ATTENTION TO THE BIBLICAL STORY OF THE PRECURSORS OR MOTHERS OF ALL BODIES IN THE UNIVERSE

- How can we rationally reconcile science and faith in a way that God deniers can plainly accept the proof of the universe's creation?
- Can atheists, rationalists, and all other freethinkers talk themselves out of denying God?
- How can we optimize the relationship between believers and atheists to better understand the origin of the universe and the existence of God?
- Which theories are especially damaging to our understanding of the universe's creation?
- What should scientists, laypeople, faculty, students, and the general public be doing when it comes to decoding the universe's origin?
- Can Christians talk themselves out of embracing secularism, while doubting God and the Biblical account of creation–don't they do that?
- How can Christians help believers accurately understand the universe's formation scientifically and beat all odds suggesting that science and the Bible are opposed?
- How can we rationally reconcile science and faith in a way that those who don't trust the Bible 100% can easily accept the proof of God creating the universe in 6 days?

Read this chapter to discover the priceless holy grail!

In fact, during the cascade of split-gathering of the bulk of the initial matter in the universe, the precursors of many galaxies were formed. About this stage, the precursor of the Milky Way Galaxy (i.e., the galaxy that the Solar System belongs to) was born, and split-gathered into the precursors of bodies including the precursor of the Solar System, which, while split-gathering, yielded the precursor of the Sun and the precursor of the bodies orbiting the Sun. After the precursor of the bodies orbiting the Sun escaped the precursor of the Sun, its fluids started flowing together before being decomposed into fluid layers of the precursors of the bodies orbiting the Sun. The layers of the precursors of the planetary systems were not just stacked one on top of the other, but they were separated by other tiny layers that were the foundation of some asteroids. For instance, between the fluid layers of the precursor of Venus and the fluid layers of the precursor of the Earth-Moon system, there were tiny fluid layers, which were the foundation of the precursor of some asteroids located between Venus and the Earth-Moon system. Likewise, between the fluid layers of the precursor of the Earth-Moon system and the fluid layers of the precursor of the Martian planetary system, there were smaller layers, which became the precursor of the asteroids between Mars and the Earth. Fig. 15 illustrates the positions of the fluid layers of some body precursors in the Solar System.

Fig. 15: Fluid layers (according to the turbulence zones) of the precursors of the bodies orbiting the Sun

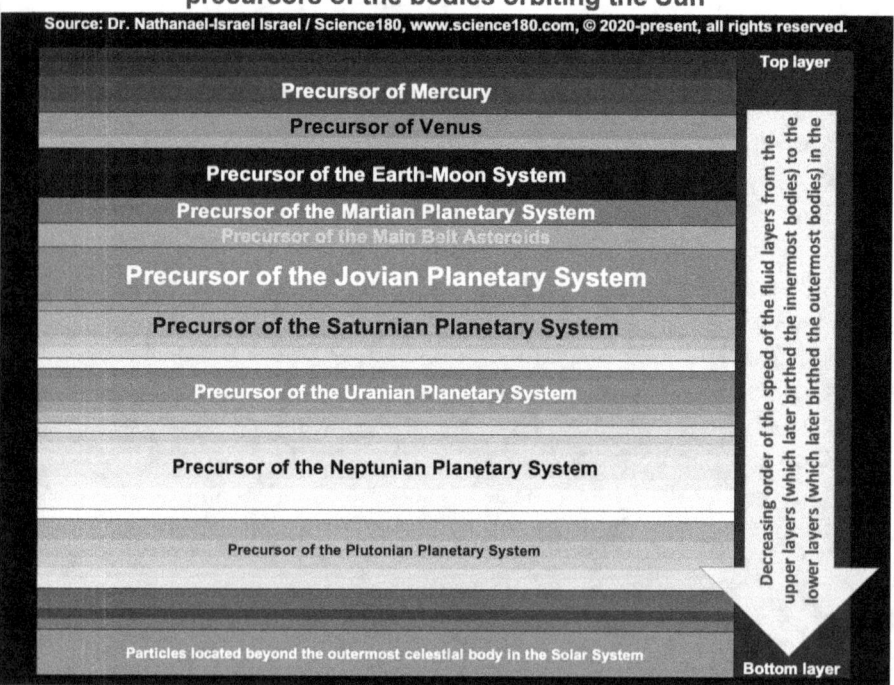

The precursors of the planetary systems in the Solar System split-gathered to yield the precursors of the primary planets and the precursors of the bodies orbiting them (satellites, rings, dust, etc.). Likewise, the precursors of the asteroid systems were split-gathered into the precursors of their primary asteroids and the precursors of the satellites, rings, and dusts around them. The precursor of the planets, asteroids, satellites, and rings went through changes and became what they are today. After the precursor of the Earth-Moon system split from the rest of the precursor of the bodies orbiting the Sun, it was split-gathered into the precursor of the Earth and the precursor of the Moon. Afterwards, each of these 2 precursors went through the "maturation" process, meaning the processes that gathered their fluid layers into a spherical shape. The precursor of the Earth was collected into the Earth. After going through some changes and traveling thousands of miles away from the Earth, the precursor of the Moon was gathered into the Moon. The formation of the Earth, as well as that of the celestial bodies with a crust, involved processes that allowed the fluid layers of the precursor to form microscopic particles that were gathered into minerals and rocks.

At this point, I would like to elaborate a little bit on the precursors of the atmosphere, medium, or space between bodies or systems of bodies in the universe. The precursor of the Earth's atmosphere is explained in Genesis 1:6-8, where the separation of waters from waters led to the formation of the sky. Likewise, most celestial bodies could have gone through a similar stage whereby a space was formed between them. For the sake of time and space, I will not elaborate much on the formation of the space, medium, and dust between galaxies, stars, and globular clusters in this book, but I have done so in *"Turbulent Origin of the Universe."* In general, the origin of the atmospheres of celestial bodies can be traced back to the separation of the fluid layers of the bodies they surround. In the case of the Earth, the precursor of the space that was formed around the precursor of the Earth-Moon system was the precursor of the immediate sky we can see from the Earth. On the scale of the Solar System, and those of the stars and other celestial bodies we can see in the sky, the space formed between their bodies altogether is the visible sky. In my book on the origin of chemical particles, I also addressed the chemical composition of these spaces.

So far, I talked about the precursors of celestial bodies, but in fact, all spiritual things and beings that God created also had a precursor. For example, angels were made of a matter that is very close to the initial matter of the universe. This can also explain why spiritual beings are able to move very fast and reach thousands and even millions of miles away in the blink of an eye. Moreover, the ability of spiritual beings to be transformed into other types of bodies can be related to the ability of the turbulent prima materia to become anything. In the chapter on the formation of angels, I provided more details. In other words, spiritual beings like angels, demons, witches, spirits involved in magic, some UFOs, and others (whose existence for now cannot be properly proven physically without using faith, for most of them are invisible), had been originally made from a precursor of matter that I labeled as the "precursor of spiritual beings and spiritual things" except God. I excluded God from this debate because I cannot dare to address His origin. Although I dare to use scientific data to explain the origin of the universe, I have never tried and I will never try to explain the origin of God by any means, for I personally believe that God has no beginning and no end. Otherwise, I will ask Him when I see him face-to-face, provided I still need to know.

If I were to summarize what I said in this chapter and the previous one, I could not fail to emphasize that scientific evidence has demonstrated the formation and separation of fluids during the formation of celestial bodies. Interestingly, just as the scientific data proved, the Bible alludes to fluid dynamics during the formation of the universe. At this point, I would like to explain how I decoded this mystery from the Biblical story of creation. Indeed, on the morning of October 12, 2018, as I was walking back home after dropping off my first daughter Josephine Gabriella-Raphae at her school, I was inspired that the story counted by Moses in Genesis 1:3 about the separation of the waters above from the waters below may be referring to the stretching and separation of precursors of the bodies surrounding the Earth or the Earth-Moon system. When I came home, I closely looked at the story of Genesis. I then noticed that, indeed, while recounting the story of creation that God revealed to him, Moses said that on the second day of creation, God created a space between some bodies of water:

> Genesis 1:3 *Then God said, "Let there be an expanse in the midst of the water! Let it be for separating water from water". 7 So God made the expanse and it separated the water that was below the expanse from the water that was over the expanse. And it happened so. 8 God called the expanse "sky". So there was evening and there was morning – a second day.* (Tree of Life Version).

This story about some events of the second day of creation has many scientific implications. It suggests that at one point, the fluids in the universe or at least the fluids of the precursor of the Earth or the fluids of the precursor of the Earth-Moon system, were near, together, or mixed with the fluids of other bodies. If the fluids that would become the Earth or the Earth-Moon system could be distinguished by Day 2, it means that, during the movement of the eddies that contained the other precursors before their separation, some movements were going on, and some agglomerations or clusters of fluids were forming and were distinguishable. The split or division of the waters suggests that a force or something was applied to separate them without destroying their constituents. Because it was on Day 3 that the waters that became the Earth were gathered, I understood that on Day 1 and even at the beginning of Day 2, the precursor of the Earth was not fully gathered into one body yet and/or all the waters over it had submerged it by the end of Day 2. The separation of the precursors of the bodies could have occurred before these bodies got their final shape. In scientific terms, it is as if the bodies took their final shape after they separated from their precursors. Because the precursor of the Earth-Moon system could have separated from its mother by the end of the 3rd day, I deduced in 2018 that it could have taken a maximum of 24 to 72 hours for the precursors of the Earth and the precursor of the Moon to break up from their common mother precursor, the precursor of the

CHAPTER 13: DISCOVER THE MOTHERS OR PRECURSORS OF THE BODIES IN THE UNIVERSE

Earth-Moon system. About 3 years before I discovered the scientific formula of the birthdate of the celestial bodies (that I will explain in the next chapter), I understood and held onto the fact that, the fluids that became the Earth and the Moon could have taken up to 24-72 hours before separating. The fact that the story in Genesis 1 talks about waters below and waters above suggested to me in 2018 that the precursor of the Earth or that of the Earth-Moon system was somewhere in the middle of a cluster of water or fluids around the second day of creation. Moreover, three years before I decoded the formula of the birthdate of the celestial bodies, I felt like some of the waters mentioned in Genesis 1:3-8 could have been:

- the precursors of Venus and/or the precursor of Venus, Mercury, the Sun, or even the precursors of other stellar systems below, above, or beyond the Solar System; or
- the precursor of the Martian planetary system and/or the precursor of other planetary systems or asteroids downstream of the precursor of the Earth.

In other words, by 2018, I felt like the waters above could have been the precursors of other stellar systems above the Solar System. In those days, although I did not know how to mathematically demonstrate these ideas, I kept thinking about what could have happened as I continued analyzing the scientific data. When I dared to narrow down what could have been the waters above and waters below the expanse mentioned in Genesis 1:3-8, I felt like the waters above could have been the precursor of Venus and the precursors of other asteroids between Earth and Venus. I also felt like the waters below could have been a precursor to the Martian planetary system and to the asteroids between Earth and Mars. It seemed to me that the story in Genesis 1:3-8 can also be referring to the split-gathering of the precursor of the Earth-Moon system into the precursor of the Moon and the precursor of the Earth. Even if those scenarios are incorrect, at least it is important to notice that it could have taken a maximum of 24 to 72 hours before the breakup of the precursor of the Earth-Moon system occurred. This interpretation I had for the Biblical story as of 2018 played a key role in my modeling of the breakup of the precursors in the Solar System.

I believe it was the breakup, shearing, and stretching of the precursors of the fluids of the celestial bodies that Moses was trying to describe in Genesis 1:3-8 where he talked about waters being separated from waters. In the Mosaic account of creation, waters were separated from waters, but if you did not know what that means, then this segment of fluid breakup in the beginning will help you! Moses wrote Genesis several thousand years ago. In those days, little was known about science and there was no space agency collecting data at that time. Most scientific terminology used in astronomy today was developed in recent centuries. Advances in fluid dynamics that allowed me to start having a glimpse at physical events like fluid stretching, breakup, or separation happened lately. Therefore, unless God, the Creator of all things revealed things to Moses through prophecy, Moses could not

Science180: First Stop for Your Universe-Creation Needs

have been able to use any human knowledge to accurately describe the formation of the Universe. The agreement between Moses' story and science proves the veracity and accuracy of the Bible and the stories it contains. We need to be careful not to take God's word lightly. The next chapter will show you how I calculated the exact date that the celestial bodies were formed and how the scientific evidence perfectly matches the creation timeline mentioned in the Bible.

Another Book by Nathanael-Israel Israel:
SCIENCE180 ACCURATE SCIENTIFIC PROOF OF GOD

THE FIRST AND THE ONLY SCIENTIFIC BOOK THAT TALKS TO ANTI-CREATIONISTS, EVOLUTIONISTS, BIG BANG PROPONENTS, ATHEISTS, AND ALL OTHER FREETHINKERS AND RATIONALISTS ABOUT THE UNIVERSE'S FORMATION, AND THEY BEG TO KNOW MORE ABOUT GOD, THE CREATOR, WHOM THEY DENY.

As you read this historic book, you will:

- Scientifically know what the one clear sign is you should always pay attention to in your efforts to decipher the primary cause and the key drivers of the fundamental processes responsible for the universe's formation.

- Discover the only way to scientifically know if God exists and, if so, which of the thousands of beings worshipped across the globe is the true God

- Accurately answer the most critical universe-origin and life-origin questions so you can stop standing in tension with consequential question marks, including those related to religion and reason or the so-called war between science and the Bible

- Discover the errors in the scientific and religious theories about the universe-origin and life-origin that are putting you at a high risk you will never recover from if you don't quickly and confidently learn how to rationally take control over threats lurking at the edge of your efforts to understand the universe and life today

- Challenge the cosmological status quo and embrace the real change that will disrupt the hidden cages that may be holding you and that you ignore

- Definitively answer all your doubts about the source or author of the universe and life ... (learn more at Science180.com/godproof)

- Understand that religion or faith, reason or science can coexist and can be properly reconciled to accurately lead you to the correct source of everything in the universe
- Satisfy your burning desire for freedom from beliefs and scientific theories about the universe's origin and life-origin that suffocate you and bind your mind, faith, unbelief, heart, and education
- Scientifically set on fire all false theories or dogmas about the existence of God, the Creator, that are enslaving humankind

Whether you are a believer, unbeliever, freethinker, administrator, politician, curriculum designer, curriculum specialist, education policymaker, teacher, librarian, school board member, researcher, parent, student, clergy, or a layperson, as long as you are really seeking to scientifically understand the rational proof of the existence of God, *"Science180 Accurate Scientific Proof of God"* is the much-admired book written for great people just like you! Grab your copy today and start reading it! Don't wait any longer!

Dr. Nathanael-Israel Israel is a Beninese-American scientist, entrepreneur, and international consultant who shows people of all ages and educational backgrounds how to scientifically decode the formation of the universe and of life, and who is acknowledged as the creator of the Chemicals Turbulent Origin Formula™, the inventor of the Life Turbulent Origin Formula™, and the discoverer of the Universe Creation Formula™. He is the Founder of Science180 Academy, which is trailblazing the reconciliation between science and the creation.

CHAPTER 14

HOW DID ALL RELIGIOUS AND SCIENTIFIC BOOKS EXCEPT THE BIBLE MISS THIS? DISCOVER WHAT THE FORMULA OF THE UNIVERSE'S BIRTHDATE IS SHOUTING AT THE BIBLICAL ACCOUNT OF CREATION! WOW, YOU WON'T BELIEVE THIS!

Is there a scientific formula that can help us to properly decode the universe's origin? I know you have heard of or you are used to the term "too good to be true," but what you may have never heard of often is "too good, and it is also very true." In this chapter, you will read a simple scientific recipe, too good and very true about how to accurately decode the timeline of the formation of celestial bodies in the universe that is going to shock you. For you have never read this before, and it is a breakthrough "they" won't teach you at any top university in the whole world. Let's get down to it now.

14.1. Why I demonstrated the timeline of the formation of the universe

The timeline of the formation of the universe (a hot topic across the globe) can be properly calculated, and in this crucial chapter, I will walk you through the scientific proofs of the exact duration of the formation of the celestial bodies. By the time you finish reading this chapter, you will fully understand how long it took for the celestial bodies to be formed, whether it was a matter of days like the Bible says or billions of years, as some theories claim. In fact, in the Bible's Book of Genesis, the creation story was recounted over 6 days. In that Biblical story, the timeline of the formation of the Earth, the Moon, and the Sun was given: the formation of the

Earth was completed on the 3rd day of creation, while that of the Moon and Sun was completed on the 4th day. But how can a human being scientifically demonstrate this timeline? This was a challenging task that most people have not spent their time working on, and no human being witnessed the creation of the world. Some famous believers even publicly said that it is impossible for science to ever prove creation.

In general, since the Biblical story of creation was recounted in the Bible, most people either just believed it or rejected it. Furthermore, the Biblical creation story also said that all living things were formed between the 3rd and the 6th day of creation—but, due to space constraints, I will not elaborate much on the formation of life in this book (but I handled that in a different book). People who deny the Bible, do not accept the Biblical timeline of creation; they don't even think that God exists. Instead, some of those unbelievers believe in the Big Bang theory, which supports that it took billions of years for the universe to be formed. Moreover, those who believe in evolutionism (advanced by Darwin) hold that life arose over billions of years through a series of processes. Even among the people who claim to believe in God, some fail to acknowledge that God created the universe and all of its content in six 24-hour days. In other words, although many believers disagree, some people think that God created the universe over billions of years. Expressed in a different way, in addition to people who deny the Bible, many people who claim to believe in the Bible support the idea according to which the universe was not formed in 6 days, but by processes that took billions of years. For years, I have wondered how someone can claim to believe in the God revealed in the Bible but still deny the 6 days of creation story recounted in the first book of the Bible. As a believer in the God revealed in the Bible, and as a scientist, I know that the Bible cannot lie, and that any interpretation deviating from the word of God must be a misinterpretation of the facts about the universe. I searched the literature, but I could not find a scientific roadmap to prove the timeline of creation. Instead, in addition to the wrong theories elaborated by unbelievers, I realized that, because most theories on intelligent design or creationism elaborated by some believers were unable to fully grasp the complexity of the factors involved in the formation of the universe, they left out the real demonstrations that are needed to prove creation, but mistakenly addressed issues irrelevant to the toughest questions that secular, unbelieving scientists are asking. Yet some believers and their leaders act as if they have all the answers about creation that they think unbelievers do not want to hear. This attitude caused many people to resist the truth, and to persist in the wrong direction.

When I considered how many lives have been lost because of how people perceive the origin and fate of the universe, and when I saw efforts being made across the globe to decode the universe, I knew that I needed to do something historic to settle the disagreement between the scientific and the biblical timelines of the formation of the universe. But again, how can I demonstrate the Biblical timeline of creation that nobody has properly, scientifically handled throughout the ages? After pondering on the question and analyzing the evidence for years, I got the answer, and because this book would have definitely been incomplete if I didn't correctly address this issue scientifically, I wrote this chapter to share some of my findings. It was a long journey filled with stories and sacrifices that I could not recount here in totality. In *"Turbulent Origin of the Universe"*, I for instance detailed my approach and findings, but here, because space is limited in this book, I will not detail the history of my discovery, but I will summarize a few critical things, hoping that those who want the details can follow up with *"Turbulent Origin of the Universe"* and also with my book on the origin of life. I know you will like this chapter, for, because it is a cornerstone chapter, I will break things down properly to ensure everybody understands it. Without any more introductions, let's start the demonstration!

From the time I started working on the origin of the universe to the time I finally discovered the formula that properly explains the duration of the formation of the celestial bodies (e.g. Earth, Moon, and Sun), about nine years elapsed. During my investigation, I realized that to determine the duration of the formation of celestial bodies, I must rely on my insight into the turbulent processes that shaped the universe. In *"Turbulent Origin of the Universe"*, I lengthily showed what I summarized earlier in this book, that during the split-gathering of the precursor of a system of celestial bodies (meaning during the process which split and gathered together the precursor of that system of bodies into precursors of primary and secondary bodies), the precursor of the secondary bodies escaped the precursor of its primary body at about the escape velocity of the primary body, then, it flowed and split-gathered into precursors of individual secondary bodies. It was like a mother birthing daughters.

After escaping the precursor of their primary body, the precursor of the secondary bodies traveled for a certain time before reaching a point where its daughter bodies were collected into celestial bodies. For instance, when the precursor of the Solar System was split-gathering, the precursor of the bodies orbiting the Sun escaped the precursor of the Sun at about the escape velocity of the Sun. Likewise, when the precursors of the planetary systems were split-gathering, the precursors of the satellites and rings escaped the precursor of their primary planet at about the escape velocity of that primary planet. For instance, when the precursor of the Earth-Moon system was split-gathering, the precursor of the Moon escaped the precursor of the Earth at about the escape velocity of the Earth, and then it flowed before organizing itself later into the Moon.

To mathematically calculate the duration of the events related to the formation of the celestial bodies in a language that everybody would understand, I invented three parameters to break down the processes involved:

- Escape time of the primary bodies
- The semi-major axis timescale of the bodies and
- Circumference timescale of the celestial bodies

After going over what each of these terms means, I will scientifically establish the exact formula of the date of birth of the celestial bodies in the universe. Then, I will compare that scientific formula to the Biblical timeline of creation. As you will see very soon, the match is remarkably astonishing, and I can't wait to show you. Let's go!

14.2. Escape time of the precursor of the primary bodies

The escape time is the amount of time required for the precursor of the secondary bodies of a system of bodies to escape the pull of the precursor of its primary body. Imagine pouring water out of a bowl. Depending on the amount of water and the rate at which it is poured out, a certain amount of time must pass before removing a certain amount of water from the bowl. When I say that the precursor of secondary bodies escaped the precursor of their primary body, look at it as the fluids of the precursor of secondary bodies being poured out of the fluids of the precursor of the primary body, or as a body of fluid escaping another body of fluid.

Just as a certain amount of time is needed for water to be poured out of a container, so also a certain amount of time was required before the fluids of the precursor of the secondary bodies could escape the precursor of their primary body. For instance, when the Solar System was being formed, it took some time before the precursor of the bodies orbiting the Sun escaped the precursor of the Sun. In the case of the celestial bodies, the amount of time required for the precursor of the secondary bodies to escape the precursor of their primary body is what I call the escape time of the precursor of the primary body. In the case of the precursor of the Solar System, the escape time of the precursor of the Sun is the time needed for the precursor of the bodies orbiting the Sun to escape the precursor of the Sun. Such a time was required for the fluids of the bodies orbiting the Sun to leave the surface of the precursor of the Sun alone. That time is about the time required for the precursor of the Sun to be fully formed and ready to be collected. Very soon, I will show you how the escape time of the precursor of the Sun played a crucial role in the duration of the formation of the Sun. Likewise, a certain amount of time was required for the precursor of a planetary system to split into the precursor of the primary planet and the precursor of the bodies (satellites and rings) orbiting it. This time was needed so the precursor of the satellites and rings of a planetary system could escape the precursor of their primary planet. This timescale is what I call the escape time of the planets' precursors. However, using data on the innermost rings of the planetary system, the semi-major axes of the innermost satellites, and the escape velocities of the planets, I have demonstrated that the escape times of the precursors of the planets in the Solar System are negligible, as they are generally less than 30 minutes. Because the Sun is about the middle of the Solar System, once the precursor of the bodies orbiting it left the precursor of the Sun, the Sun just started swirling to form. In other words, once the escape time of the precursor of the Sun had elapsed, the latter was ready to collect all its fluid together to form the Sun.

As of the publication of this book, Mercury is known as the most renowned innermost body in the Solar System. In other words, Mercury is the closest body to the Sun. Because of the organization of the fluid layers of the precursor of the bodies orbiting the Sun, by the time the precursor of Mercury reached the semi-major axis of Mercury (meaning the distance separating Mercury and the Sun) and escaped the stack of fluid layers of the bodies orbiting the Sun, the precursor of all the bodies orbiting the Sun must have also reached that position. The fluid layers of the precursor of Mercury were on top of the fluid layers of the bodies orbiting the Sun. Because the precursor of the bodies orbiting the Sun escaped the precursor of the Sun at about the escape velocity of the Sun, the duration of time it took for the precursor of the bodies orbiting the Sun to reach the position of Mercury is about the semi-major axis of Mercury divided by the escape velocity of the Sun. I showed that the time it took for all the bodies orbiting the Sun to escape the Sun was about the duration of time I just said. In other words, the escape time of the precursor of the Sun is about the semi-major axis of Mercury divided by the escape velocity of Mercury.

Escape time of the precursor of the Sun = Semi major axis of Mercury / Escape velocity of the Sun

In my book called *"From Science to Bible's Conclusions,"* I provided more details. NASA and other space agencies have determined that Mercury's semi-major axis is 57,910,000 km and that the Sun's escape velocity is 617.6 km/s. Therefore, the escape time of the precursor of the Sun is about 26.046 hours (meaning 57,910,000 km divided by 617.6 km/s). This is to say that it took about 26.046 hours for the fluid layers of the bodies orbiting the Sun to escape the precursor of the Sun so the precursor of the Sun could be ready to swirl and form the Sun.

After escaping the precursor of the Sun, the precursor of the planets, asteroids, and satellites traveled for some time before being collected into their corresponding celestial bodies. Likewise, after splitting from the precursor of their primary planet, the precursor of the satellites journeyed for some time before being collected into celestial bodies. In general, after splitting from the precursor of their primary bodies, the precursor of the secondary bodies traveled for a certain distance before being wrapped around into the corresponding celestial bodies. Below, I will explain the duration of those travels.

14.3. Semi-major axis timescale of the secondary bodies

The semi-major axis timescale is a term I coined to express the duration of time the precursor of a secondary body traveled from about the position of its primary body to the position of its orbit before being collected into a celestial body. For example, the semi-major axis timescale of a planet in the Solar System is about the time that

Nathanael-Israel Israel: Has had the Honor to be Acknowledged the First Human Being that Scientifically Reconciled Science & the Biblical Creation the precursor of that planet traveled away from the precursor of the Sun before being collected into a planet. Likewise, the semi-major axis timescale of a satellite is about the time that the precursor of that satellite traveled from the precursor of its primary planet (after escaping it) until the time the precursor of that satellite was ready to collect into the satellite. The estimation of that timescale requires an intimate understanding of the distance and speed of that travel. To shorten a long demonstration, I showed (for instance, in my book called *"From Science to Bible's Conclusions"*) that the distance traveled (with respect to their primary body) by the precursor of the planets, asteroids, and satellites before being collected into celestial bodies is about their semi-major axis, meaning the average distance between these bodies and their primary body. For instance, the time it took for the precursor of a planet in the Solar System to travel from the precursor's position to where it was collected into a planet is about the distance separating that planet from the Sun divided by the Sun's escape velocity. Because the distance separating the planet from the Sun is called the semi-major axis of the planet, the time elapsed for the precursor of a planet in the Solar System to escape the precursor of the Sun and reach the orbit of that planet is about the semi-major axis of that planet divided by the escape velocity of the Sun. I divided the semi-major axis (the distance between the Sun and a planet) by the escape velocity of the Sun (a speed of escape) because, like I already showed, the precursors of the planets escaped the precursor of the Sun at the escape velocity of the Sun and traveled at about that speed until reaching the point where they separated from the stack of fluids carrying them. This duration of time is what I labeled the semi-major axis timescale of the planet. In the case of the Earth, the semi-major axis timescale of the Earth is equal to the semi-major axis of the Earth (which is the average distance between the Earth and the Sun) divided by the escape velocity of the Sun (which is the speed at which the precursor of the Earth, which was embedded in the precursor of all the bodies orbiting the Sun, escaped the precursor of the Sun).

Semi major axis timescale of the Earth = Semi major axis of the Earth / Escape velocity of the Sun

NASA (2018) has shown that the semi-major axis of the Earth is 149,600,000 km. Therefore, considering the escape velocity of the Sun (617.6 km/s), the semi-major axis timescale of the Earth is about 67.286 hours (149,600,000 km divided by 617.6 km/s). This means that about 67.286 hours after the precursor of the bodies orbiting the Sun escaped the precursor of the Sun at about 617.6 km/s, the precursor of the Earth was formed and ready to swirl to form the Earth. In reality, before the precursor of the Earth was formed, the precursor of the Earth-Moon system was first formed and split from the stack of fluid layers of the bodies orbiting the Sun. I showed that, after its formation, the precursor of the Earth-Moon system split almost immediately into the precursors of the Earth and the Moon. In other words, about 67.286 hours after the beginning, the precursor of the

Earth-Moon system was formed and split into the precursor of the Earth and the precursor of the Moon. For more details on the demonstrations, please consult *"Turbulent Origin of the Universe."* Later, I will use this formula to address the duration of time of the Earth.

Because the precursor of the satellites of the planets escaped the precursor of their primary planet at about the escape velocity of the planet and traveled at about that speed for about the semi-major axis of a satellite (which is the average distance separating the satellite from its primary planet), the time elapsed since the precursor of a satellite escaped the precursor of its planet to reach the orbit of that satellite is about the semi-major axis of that satellite divided by the escape velocity of the primary planet. This duration is what I call the satellite's semi-major axis timescale. Applying this formula to the Moon, the duration of time elapsed from the moment the precursor of the Moon escaped the precursor of the Earth (at about the escape velocity of the Earth) until the precursor of the Moon was about to swirl to become the Moon is about the semi-major axis of the Moon (which is the distance separating the Moon and the Earth) divided by the escape velocity of the Earth (which is about the speed at which the precursor of the Moon traveled away from the precursor of the Earth after their split). This duration of time is what I call the semi-major axis timescale of the Moon.

Semi major axis timescale of the Moon = Semi major axis of the Moon / Escape velocity of the Earth

Because the semi-major axis of the Moon is 384,400 km and the escape velocity of the Earth is 11.186 km/s, the semi-major axis timescale of the Moon is 9.546 hours (meaning 384,400 km divided by 11.186 km/s). In other words, after escaping the precursor of the Earth at about 11.186 km/s, the precursor of the Moon traveled for about 9.546 hours before reaching a point when its fluid layers could swirl and be gathered together into the Moon. I will shortly show you how this timescale contributed to the duration of the formation of the Moon.

14.4. Circumference timescale of the celestial bodies

The circumference of a circle is the enclosing boundary of that circle or the distance around that circle. The circumference of a spherical object (think of a sphere like a body shaped like an orange) is the length of its perimeter or rim at the equator. Celestial bodies like the Sun and the planets in the Solar System have a spherical shape. Because the precursors of these celestial bodies were fluid layers, it took some time for these fluid layers to be collected together into spherical bodies. In *"Turbulent Origin of the Universe,"* I detailed how this gathering together of fluid layers occurred. Here, I will just summarize that process, hoping that those who want the long demonstration can consult the aforementioned book.

In short, as the fluid layers of the precursor were journeying to the position where they would be collected as a whole, various structures (e.g. vortices) formed in them and amalgamated or merged into larger and larger structures as they were moved by a turbulent fluid flow. In *"Turbulent Origin of the Universe"*, I lengthily showed that, once the precursor of a celestial body was formed, and reached about the position of its orbit, its fluid layers were as long as about the circumference of the body it would form. Then, that precursor swirled at about the orbital speed of the bodies it would form. For instance, after the precursor of the Sun was formed, the length of its fluid layers was about the circumference of the Sun; then these fluid layers swirled at about the orbital speed of the Sun to form the Sun. Likewise, at the time the precursor of the Earth was about to collect into the Earth, the length of its fluid layers was about the circumference of the Earth. Then, these fluid layers swirled about the orbital speed of the Earth to form the Earth. In the same manner, when the precursor of the Moon reached about the orbit of the Moon and was about to gather into the Moon, its fluid layers were as long as about the circumference of the Moon. Furthermore, these fluid layers swirled or wrapped around at about the orbital speed of the Moon to form the Moon.

The circumference timescale of a celestial body is a term I invented to describe the time required for the fluid layers of its precursor (which reached the position or orbit of its daughter bodies) to be gathered together into a celestial body. Because of the length of the fluid layers at the last stage, their linear conformation was about the circumference of their daughter bodies, and because they swirled at about the orbital speed of their daughter bodies, the duration of their swirling was about the circumference divided by the orbital speed of their body. In other words, the circumference timescale is the circumference divided by the orbital speed of the body.

To ensure that all readers of this book understand the meaning of the circumference timescale and how I calculated it, let me break this formula down a little more. Imagine a cord or thread whose length is "L" that you want to roll to make a circle or to roll around a ball whose circumference is "L". If you wind the entire thread to form a circle, the circumference (i.e., enclosing boundary or the distance around) of that circle will be L, the length of that cord. Imagine you rolled that thread at a speed of V. The duration of time it will take to make that full circle would be the length of that cord divided by the speed used to form it. In other words, knowing the circumference of a circle and the speed at which it was formed by rolling a cord or a linear rope, the time elapsed to form that circle can be estimated. I am pretty sure you have already done a similar calculation regarding the time you took to travel a certain distance. For instance, if you drive 40 miles at 10 miles per hour, it will take 4 hours, calculated by dividing 40 miles by 10 miles/hour. In other words, time is calculated by dividing a distance by a speed. This is also true for the time spent traveling all around a spherical body at a certain speed. This difficult part (for some people) here may be the formula of the circumference of a body knowing its radius (R):

$$\text{Circumference} = 2\pi R = 2 \times 3.14 \times R$$

The same thing I explained above can be done for a spherical body. The circumference at the equator of a spherical body is like the maximum distance that can be traveled all the way around that body. In other words, if you take a sphere, and using a cord in your hand, you start measuring the circumference all the way around from the equator all the way to the pole, you will notice that the circumference decreases. For instance, a slice of an orange in the middle is larger or longer than a slice toward its pole. Putting this statement another way, the circumference at the equator or at the middle of a sphere is the maximum distance that can be traveled all the way around that sphere. If you peel a tiny slice from the equator of an orange, open it up, and lay it flat to form a cord or line, you will notice that the length of that slice can be roughly equal to the circumference. If you try to roll or curl that slice to make a circle, as I explained above, the time it will take to do it will be equal to the length of that slice (which was the circumference of the orange at its equator) divided by the speed of the curling motion. To make it clearer, the time it could take to move around the circumference of a sphere is equal to its circumference divided by the speed of the motion. Everything I explained in this segment so far can also be applied to celestial bodies by saying that, the time it will take to travel their circumference all the way around is equal to that circumference divided by the speed of the travel. I hope you get it!

I have showed that, at one point, the precursors of the celestial bodies were fluid layers, which ended up being spiraled or whirled around their axis to form celestial bodies, some of which are spherical. The rapid swirling of the fluid layers of the precursors can also explain the overall flatness of the poles and the raising up of the equator, which is like the midway point between the north and south poles. I showed that the speed of the spiraling or swirling of the fluids of the precursor of these bodies was about the orbital speed of their bodies. Therefore, the time it took for the fluid layers of the precursor of a celestial body (whose length was about the circumference of that celestial body) to roll at about the orbital speed was about the circumference of that celestial body divided by its orbital speed. I will recapitulate the circumference timescale of the Earth, the Moon, and the Sun with the following formula:

Circumference timescale of the Earth = 2 x 3.14 x (Radius of the Earth) / (Orbital speed of the Earth)

Circumference timescale of the Moon = 2 x 3.14 x (Radius of the Moon) / (Orbital speed of the Moon)

Circumference timescale of the Sun = 2 x 3.14 x (Radius of the Sun) / (Orbital speed of the Sun

Using the above formula, I will now calculate the circumference timescale of the Earth, the Moon, and the Sun. Indeed, NASA (2018) already established that the radius of the Sun is 696,000 km, that of the Earth is 6,378.14 km, and that of the Moon is 1,738.1 km. NASA also established that the orbital speed of the Sun is 19.4 km/s, that of the Earth is 29.78 km/s, and that of the Moon is 1.02 km/s. Using these values and the formula I established above for the circumference timescale, I showed that:

- the circumference timescale of the Earth is 0.374 hours, meaning 22.42 minutes;
- the circumference timescale of the Moon is 2.96 hours; and
- the circumference timescale of the Sun is 62.584 hours.

This data indicates that, after the Earth's precursor formed, it took about 22.42 minutes for its fluid layers to collect into the Earth. Likewise, about 2.96 hours after the Moon's precursor reached the Moon's orbit, its swirling led to the Moon's formation.

The circumference timescale of about 24% of the bodies I studied in the Solar System is less than 1 second. In general, about 98% of the bodies I studied have a circumference timescale less than 1 hour. Besides the Sun and the Moon, only 5 bodies have a circumference timescale higher than 1 hour. I provided more details in *"Turbulent Origin of the Universe."* In short, the fluid layers of the precursors of the celestial bodies in the Solar System were collected in a very short amount of time after these precursors were formed.

14.5. Birthdate of celestial bodies

I defined the birthdate of a celestial body as the time elapsed from the beginning of the formation of the universe until the formation of that body was completed. Considering the demonstration I did so far in this chapter, the duration of time of the formation of the Sun involved two components: (1) the time needed for the precursor of the bodies orbiting the Sun to escape the precursor of the Sun and (2) the time required for the precursor of the Sun to be collected together. Expressed in a different way, the duration of the formation of the Sun is the sum of (1) the escape time of the precursor of the Sun and (2) the circumference timescale of the Sun.

Because the planets and satellites have a semi-major axis (i.e., the distance separating them from their primary star), their precursors traveled for a certain distance before being able to swirl and form those bodies. In previous segments, I showed that the precursors of the planets traveled for about their semi-major axis away from their primary body before reaching a point where they were collected and the duration of that travel was what I called the semi-major axis timescale of the planets, a time that elapsed before the swirling corresponding to the circumference timescale of the planets occurred. In other words, the duration of the formation of a planet in the Solar System is the sum of (1) the duration of the travel of its precursors from the precursor of the Sun to the orbit of that planet and (2) the duration of the collection of the fluid layers of the planet once they reached the semi-major axis of that planet. Similarly, the duration of the formation of the satellites took into account (1) the semi-major axis timescale of the satellites and (2) the circumference timescale. But because the semi-major axis timescale of the primary planet had elapsed before the precursor of the satellites were formed, to calculate the duration of the formation of the satellites with respect to the star in their stellar system, the semi-major axis timescale of the planet must also be considered. Imagine your parents were born in a certain city and had to travel a certain distance before giving birth to you. And then, after you were born, you too traveled for a certain distance before settling somewhere and birthing your own children. The distance separating where your children were born and where your parents started their journey is the sum of the distance traveled by your parent before birthing you and the distance traveled by you before birthing your children. In other words, imagine your parent was living in a certain place and had to travel for a certain time to another place before birthing you, and then you too had to move out and travel for a certain amount of time to another place before birthing your children. The total time separating where your parent started its journey and where your children were born must include the time your parent traveled before birthing you and the time you traveled before birthing your children. The same math applies if the time elapsed can be considered: they must be added to find the total. Hence, the semi-major axis timescale of a satellite with respect to the primary star of a stellar system is the sum of (1) the semi-major axis timescale of the primary planet and (2) the semi-major axis timescale of the satellite. In the end, with respect to the primary star, the duration of the formation of a satellite is the sum of (1) the semi-major axis timescale of the primary planet, and (2) the semi-major axis timescale of the satellite, and (3) the circumference timescale of the satellite. Here is the summary of what I said so far about the birthdate as it applies to some bodies in the Solar System:

- **Birthdate of the Sun = (Escape time of the precursor of the Sun) + (Circumference timescale of the Sun)**

- **Birthdate of the Earth = (Semi major axis timescale of the Earth) + (Circumference timescale of the Earth)**

- **Birthdate of the Moon = (Semi major axis timescale of the Earth) + (Semi major axis timescale of the Moon) + (Circumference timescale of the Moon)**

Replacing the above timescales by their expressions I demonstrated earlier, the formula of the birthdate of the Earth, the Moon, and the Sun with respect to the beginning is:

- **Birthdate of the Earth = (Semi major axis of the Earth) / (Escape velocity of the Sun) + 2 x 3.14 x (Radius of the Earth) / (Orbital speed of the Earth)**

- **Birthdate of the Moon = (Semi major axis of the Earth) / (Escape velocity of the Sun) + (Semi major axis of the Moon) / (Escape velocity of the Earth) + 2 x 3.14 x (Radius of the Moon) / (Orbital speed of the Moon)**

- **Birthdate of the Sun = (Semi major axis of Mercury) / (Escape velocity of the Sun) + 2 x 3.14 x (Radius of the Sun) / (Orbital speed of the Sun)**

Considering the values that I already presented for the semi-major axis of the Earth (149,600,000 km), the semi-major axis of the Moon (384,400 km), the semi-major axis of Mercury (57,910,000 km), the escape velocity of the Sun (617.6 km/s), the escape velocity of the Earth (11.186 km/s), the radius of the Sun (696,000 km), the radius of the Earth (6,378.14 km), the radius of the Moon (1,738.1 km), the orbital speed of the Earth (29.78 km/s), and the orbital speed of the Moon (1.02 km/s), I showed (using the above formula) that:

- the birthdate of the Earth is 67.659 hours, which is equivalent to 2.819 days
- the birthdate of the Moon is 79.795 hours, meaning 3.325 days
- the birthdate of the Sun is 88.63 hours, meaning 3.693 days

In other words, based on the scientific facts, I proved for the first time in history that the Earth was born 2.82 days, the Moon 3.32 days, and the Sun 3.693 days after the beginning of the formation of the Solar System, which I showed is almost the same time as the beginning of the formation of the universe. In *"Turbulent Origin of the Universe,"* I detailed the birthdate of the other bodies in the universe. After I demonstrated the duration of the birthdate of the celestial bodies, I also proved in which month and day of the week they were born. I initially wrote many pages about this subject in the first draft of this book, but seeing the size of this book, I moved most of the writing on the date and month of creation to the book I wrote on the age of the universe.

14.6. Perfect match between the scientific timeline and the biblical timeline of the formation of the universe

The scientific demonstration I did above for the birthdate of the celestial bodies perfectly matches the Biblical story of creation. Indeed, I showed that:

- the Earth was formed 2.82 days after the beginning, meaning on the 3rd day of creation;
- the Moon was formed 3.32 days after the beginning, meaning on the 4th day of creation; and
- the Sun was formed 3.69 days after the beginning, meaning on the 4th day of creation.

Those dates are exactly what the Bible's Book of Genesis said:

Genesis 1:1 *In the beginning God created the heaven and the earth. 2 And the earth was without form, and void; and darkness was upon the face of the deep. And the Spirit of God moved upon the face of the waters. 3 And God said, let there be light: and there was light. 4 And God saw the light, that it was good: and God divided the light from the darkness. 5 And God called the light Day, and the darkness he called Night. And the evening and the morning were the first day. 6 And God said, let there be a firmament in the midst of the waters, and **let it divide the waters from the waters.** 7 And God made the firmament, and **divided the waters which were under the firmament from the waters which were above the firmament**: and it was so. 8 And God called the firmament Heaven. And the evening and the morning were the second day. 9 And God said, let the waters under the heaven be **gathered together unto one place, and let the dry land appear**: and it was so. 10 And **God called the dry land Earth**; and the gathering together of the waters called the Seas: and God saw that it was good. 11 … 12 … and God saw that it was good. 13 And the evening and the morning were **the third day**. 14 And God said, let there be lights in the firmament of the heaven to divide the day from the night; and let them be for signs, and for seasons, and for days, and years: 15 And let them be for lights in the firmament of the heaven to give light upon the earth: and it was so. 16 And God made two great lights; the **greater light** [Sun] to rule the day, and the **lesser light** [Moon] to rule the night: he made the stars also. 17 And God set them in the firmament of the heaven to give light upon the earth, 18 And to rule over the day and over the night, and to divide the light from the darkness: and God saw that it was good. 19 And the evening and the morning were **the fourth day*** (King James Version).

In *"From Science to Bible's Conclusions"* and in *"Turbulent Origin of the Universe"*, I spent pages commenting on the above verses. By consulting Hebrew, I decoded hidden mysteries in the story and underlined some of the challenges the Biblical translators faced in rendering the original Hebrew texts into English and other languages. For instance, earlier in this book, I showed (as many theologians have also done) that the "heavens and earth" mentioned in Genesis 1:1 alludes to the precursors of the heavens and the earth, not to the complete Earth and heavens known today. In fact, the Genesis story also explained that, on the first day, the

Earth was not fully formed yet, but that waters had to be divided from waters on the 2nd day (Genesis 1:6-8) and then gathered together on the 3rd day before the Earth could be formed. The scientific evidence that I presented earlier also showed that, in the Solar System for instance, fluids of the precursors of bodies started separating by the second day. In the segment I wrote about "the escape time," I showed that, about 26.046 hours after the beginning, the fluid layers of the precursors of the bodies orbiting the Sun started splitting from one another.

In the days of Moses (about 3500 years ago), modern science was not even born yet, and the terminology of fluids was not even created yet. Because the Bible was written before the contemporary terminology (including "fluids") was invented, the term "waters" used in the Biblical story of creation can also be understood as "fluids." In the days of Moses, if someone was about to talk about the separation of the fluid layers above the precursor of the Earth-Moon system, that person could say that, on the second day of creation, waters [fluids] were separated from one another until the waters [fluids] of the precursor of the Earth-Moon system were reached, and then gathered into the Earth and the Moon accordingly. Just like the scientific evidence, the Bible also said that the waters (fluids) of the precursor of the Earth were gathered together on the 3rd day to form the Earth (Genesis 1:9-13). The Biblical story also says that the Moon and the Sun were formed on the 4th day (Genesis 1:14-19).

Furthermore, as I amply demonstrated in my book *"Turbulent Origin of the Universe"*, the same mathematics that allowed me to calculate the birthdate of the Earth, the Moon, and the Sun (just as the Bible says) also proved that some celestial bodies were born several days, even months after the beginning. In other words, although my findings proved that the Biblical account of the creation of celestial bodies is correct, it is important to warn people not to think that the 6 days of creation mentioned in the Bible refer to the formation of all the celestial bodies in the universe today. In other words, although the precursors of all celestial bodies were born and going through changes before the end of the 6th day of creation, it took many more days and even months before the precursors of some celestial bodies could reach their orbit and be fully formed. This means that Genesis 1:1 is still very correct, as, at the end of Day 7, other celestial bodies were still being formed, while their precursors had already formed. It is like how a baby is conceived, stays in the mother's womb for about 9 months, and then grows through baby, toddler, teenager, young adult, old, etc. Therefore, those who think that God formed EVERYTHING in the universe in 6 days are not completely correct. For instance, many people alive today were not born during the 6 days of creation, but many thousands of years later, yet they are also creatures of God. Furthermore, because the seed of every human being was in Adam (which I addressed in another book) when he was created on the 6th day, it is not wrong to say that every human being was created on the 6th day of creation, provided we agree that the human race that Adam was carrying (as germ line) was living inside of him in a different form than what it would be after the corresponding human beings were born. Similarly, by the end of the 6th day of creation, the seed, or precursor, of every celestial body in the universe had already been created, but existed in a form different from what its daughter bodies would be after that seed underwent birthing and growth. As of today, the current generation of living things (not just human beings) is carrying the seeds of future generations. But all those seeds will be born before the end comes. Similarly, new chemicals and various forms of life are born today using particles already created by the end of the first 6 days.

Among all the religions in the world, only the creation story recounted in the Bible matches the scientific evidence. In *"Turbulent Origin of the Universe,"* I provided ample details encoded in the Genesis story of creation. But for the sake of space, I will not delve into those details here. Below, I will present a glimpse at the significance of my discoveries about creation.

14.7. Importance of the match between the scientific evidence and the Biblical story of creation

You have heard people say that God created the universe in 6 days, while others claim that the Big Bang did so over billions of years. But before my discovery (see www.Israel120.com/books), you had never heard of a human being who mathematically demonstrated the Biblical account of creation to be accurate, using the great scientific raw data hidden within wrong scientific theories. Indeed, after focusing on scientifically cracking some of the world's most unsolved problems, I discovered that, the reason most believers have not scientifically understood the creation of the universe is that they have spent years reading the Genesis story, and have neglected to properly ponder on ancient revelations about the origin of the universe and life but have learned nearly nothing about how to properly interpret it using the scientific data without checking either their faith or their mind at the door... The result is that most believers learned to study and memorize all kinds of Biblical verses and metaphorical interpretations of creation, while they trash secular theories with all their amazing raw data, and they never learned how to unconventionally acknowledge the fact that the biblical account of creation is a literal and chronological history that they are struggling to rationally decode but that can be explained scientifically without compromise... so they can have key scientific raw data (denuded of the secular theories wrapping and warping them) to properly work for and lead them to the demonstration of the conclusion of the Bible rather than starting with that accurate conclusion that, unfortunately, usually chases away rationalists and skeptics: God really created the universe in six 24-hours literal consecutive days indeed.

How can believers abandon wrong creationist theories if they think it is impossible for science to literally prove the Genesis story, or if they think that science is evil and diametrically opposed to faith, or if they continue to compromisingly embrace scientific theories that totally contradict the Bible as if people can use their minds to help God by alleging that the Biblical account of creation, written before the scientific era, lied or is inaccurate? How can the skeptic believe in the Biblical narrative of creation if most believers can't even properly prove it scientifically?

Lucky you, that is where I came in to reanalyze the raw scientific data from an original perspective to unlock the ultimate and only solution that believers and skeptics have been looking for throughout the ages…

Unlike any of the other books you have ever read on the origin of the universe, this book will scientifically show you for the first time in history how the Earth was formed in 2.819 days, the Moon in 3.325 days, and the Sun in 3.693 days. By the time you finish reading this mind-blowing masterpiece, you will know for sure that God really created the universe. As you keep reading this book, you will improve your life nonstop.

This book would have missed the point if I had not elaborated on the significance or implications of the discoveries I presented in this chapter. Indeed, I showed how the scientific evidence supports the creation of the universe in a matter of days rather than millions or billions of years, as some theories mistakenly postulate. In the Biblical creation story, the Bible clearly says that God is the Creator. Moses, who is considered the author of the Genesis creation story (that he heard from God), also wrote other books in which he repeatedly revealed that the God of Israel is the Creator of the universe.

In my books, *"From Science to Bible's Conclusions"* and *"Science180 Accurate Scientific Proof of God,"* I devoted numerous pages to the significance of the match between scientific data and the Biblical story of creation. For instance, in that book, I expounded on the history of science and how the Book of Genesis was written thousands of years before the foundations of modern science were laid. Yet, the timeline of the Biblical creation story is so accurate. I highly recommend you consult the aforementioned books for their outstanding details. Due to the limited space available here, I will not repeat those remarks, but I will instead say a few words about the precision of the calibration of the process that formed the universe.

One question worth asking is why God chose to create the world in 6 days. I will offer here my thought as it crossed my mind on November 28, 2020. Indeed, it appeared to me that if God wanted, He could have created the whole world in a few seconds, but if He had done so, the world could have been set on another tune, movement, and functionality. For the timing of the processes involved in the formation of the universe affected the functioning of the current universe. For instance, it was no accident that the formation of the Sun was completed on Day 4. The events involved in the formation of the Solar System calibrated the current characteristics of the Sun and the other bodies in the Solar System, including the Earth, the Moon, and all the chemical particles they contain. Knowing that the Sun and the Moon played a key role in the timing of events on Earth and even in the synchronization of the heavenly calendar with the earthly calendar (e.g. timing of the feasts of the Lord, whose details you can see in my book on the age of the universe), I perceived that the timing and functioning of things in the universe would have been different if God had created the universe in fewer or more than 6 days. In other words, for human beings to reflect the image and likeness of God, and for certain earthly things to be a template of some heavenly things, and for God to rest on the 7th day of creation and then set the Sabbath as a weekly template of what the 1000 years of the Millennial reign represent for the 6000 years from the creation to the rapture, the universe must have been created in 6 days. Another way of addressing the duration of creation is that, the 6 days in which God worked to create the universe were perfectly aligned with God's eternal plan and with God Himself who intended to create His own representative or ambassador on Earth: the human beings, which regretfully failed, but which God redeemed using laws associated with the world He created. Once God created human beings and beforehand everything they needs (e.g., Earth, Moon, Sun, all other organisms on Earth) and the angels that will minister to them, God declared his creatures or creation good and complete, but in fact, creation continued until today. Those who ignore this mystery think that EVERYTHING was fully formed in 6 literal days.

Creating things is not just the problem; what purpose are they created for? God decided to create a world whose functioning must be calibrated with the things and beings it would host. Hence, the timing of everything was adjusted accordingly, leaving no room for chance. That is why God had to declare the end before the beginning. Otherwise, He could have started creating a world that He would later realize was unfit for the end He intended.

Tuning the beginning of the universe to deliver the end that God expected made the development stages it would go through to appear as if everything was set in stone by God, but in fact, human beings and angels could have made certain things to go differently. Although, through His predestination, prescience, omnipotence, and omnipresence, God can control everything today, He did not choose to dictate everything in the current universe, particularly things on Earth. God's foreknowledge does not mean that He makes people do things against their will. That is why human beings and angels are still responsible for the consequences of their choices, including accepting and believing in the Savior. Nevertheless, God can sometimes impose His will on certain things and human beings so that His universal plan, which creatures are not always aware of, could stand just as planned before anything was created. Hence God is the Alpha and the Omega (the beginning and the end)! With this in mind, have you accepted the Savior? Do you believe in Him?

While some people credit God for the order in the universe, others see chaos and randomness. As a believer with a specific mission to reveal the origin of the universe to humankind, I could not find any other source or origin of the universe than the omniscient and omnipotent God, who planned and executed His design, evident in both the nonliving and living things in the universe. The harmonics within and between the constituents of the universe are so well thought out that it is unthoughtful to leave their cause to pure chance. If God's plan was not properly executed during the formation of the universe, a little change in the mother of all turbulences that shaped the origin of the universe could have sufficed to change the distribution, characteristics, and functioning of all matters in the universe, and consequently the organization of life. Therefore, it is imperative for human beings to acknowledge and praise God for His great design that established order in the universe, where the habitable Earth hosts human beings, including the unbelievers, who unfortunately do not obey God the creator, but seek their origin in a vain and useless pursuit of knowledge, not knowing that they are preparing themselves for the eternal punishment reserved for those who reject God and His word.

Meaningful of where I am coming from, I know that, if the foundations of my thoughts were not in the Bible, I would have missed many details in the demonstration of the creation timeline. For without a proper guide or compass, no one could imagine the ramifications of the events surrounding the origin of the world. But by setting my mind on the Creator and by learning from my challenging life's journey, it was easier for me to reject certain thinking pathways that could have blinded me and complicated my understanding of the origin and functioning of the universe. Therefore, although significant advances have been made in science, no scientist before me has been able to properly link existing scientific and philosophical facts to the real story of the universe's origin.

Although this complicated task could have been impossible some centuries ago, for some crucial scientific data were not collected yet, I felt like scientists could have laid the foundation for a better theory of our origin a long time ago if they were willing to refer to God and trust His account of the beginning, and not blindly accept everything in the scientific methods (which have been claiming to not be seeking truth). Moreover, some of the details I unraveled in this book may never have been discovered through laboratory experiments, and others may not have been properly understood without an intimate grasp of advances in fluid dynamics and many other scientific disciplines. A detailed insight into the origin of the universe requires an approach more comprehensive than the traditional, mere, linear laboratory experimental methods could allow. For instance, as of 2020, the largest wind tunnel used to study turbulence (a phenomenon crucial to understanding the universe) at large scales is less than 100 meters. The small size of lab equipment also suggests that it may be impossible for scientists to conduct large-scale experiments, which can allow them to clearly see and study some turbulence details. In other words, scientists may need to design larger-scale experiments than they can currently handle. Unfortunately, the cost, collection, and analysis of bigger experimental data are other obstacles facing turbulence researchers. This also means that scientists are limited in their ability to replicate large-scale turbulence found in nature. Running larger-scale experiments is like simulating aspects of the universe's formation with limited human power. Consequently, the accuracy of scientific modeling is limited, as many details are missing. I give glory to God who allowed me to discover these things.

In short, I proved in this chapter that the scientific data support the six 24-hour days of the creation story narrated in the Bible, but it was human misunderstanding that caused them not to accept or be able to demonstrate the Biblical timeline until now. Because the things that the Bible said to have occurred on the first, second, third, and 4th day of creation are now scientifically proven to be 100% true, we need to also accept the remaining story of the other 2 days of the 6 days of creation; else we are running a big risk whose price will be costlier than any funds the anti-creationist organizations may offer to say the opposite. As the person who, for the first time in history, demonstrated the timeline of creation and settled the disagreement between the scientific data and the Biblical account of creation, I must also confess that the only reliable and undeniable eternal reference that can help human beings understand how the universe began and what is awaiting it is the Word of God as expressed in the Bible. Because the universe was created by a God who declared the end from the beginning, it was founded with a clear future, and people need to stop taking the word of God lightly. Do you believe in God the Creator?

'Science180 Academy' Success Strategy:
SCIENCE180 BOOKS THAT WILL HELP YOU!

I, Nathanael-Israel Israel, broke down my discovery of the formation of the universe into many books so that you, the readers, can pick the ones that best suit your needs and interests without disappointing you or wasting your precious time. These books come in many versions (e.g., scientific version, public version, chemical version, biological version, biblical or prophetic version, pseudepigraphic version, and a children's version) targeting people according to their expertise, educational background, and interests as briefed below:

1. "*TURBULENT ORIGIN OF THE UNIVERSE*" (This is the scientific version of my book tailored to scientists and anyone interested in the detailed scientific demonstration of the universe's formation). In this book, I used the "mother of all turbulences" to scientifically demonstrate the formation of the universe so that scientists can understand and reorient the course of their research, teaching, and publishing, and accept the truth to better live today and forever. Get "*Turbulent Origin of the Universe*" today to begin an incredible journey of accurately decoding the universe and changing your life forever! Learn more at Science180.com/scientific

2. "*RECONCILING SCIENCE AND CREATION ACCURATELY*" (this is the book that I called the "Biblical or prophetic version of my book on the universe's origin, and it targets Christians and anyone interested in knowing the Biblical perspective of the creation of the universe). This important book accurately demonstrates the marvelous creation and formation of the universe

by God in six consecutive 24-hour days and answers many questions about the universe's creation so that after acknowledging Him (who deserves all the glory now and forever), human beings can choose life and avoid the terrible judgment awaiting the unbelievers in the world to come. Get this thoughtful book now to figure out what happened at the beginning, what is coming up, and why it is time to urgently rethink everything you have been told about the universe's origin so you don't eventually regret it! Don't say I did not tell you! Learn more at Science180.com/biblical

3. **"TURBULENT ORIGIN OF CHEMICAL PARTICLES"** (Called the "chemical version" of my book on the universe's origin, this elegant book targets chemists, biochemists, and anyone interested in chemistry). With *"Turbulent Origin of Chemical Particles"*, the accurate deciphering and understanding of the formation of chemicals has never been profitable and easy. Hence, this great book is THE ultimate how-to guide for great people who want to correctly decode the origins of chemicals and positively transform their lives. Get this celebrated book today. Learn more at Science180.com/chemical

4. **"ORIGIN OF THE SPIRITUAL WORLD"** (This book is what I call the pseudepigraphic or hidden version of my books on the universe's origin, and it is meant for believers who want to tap into a higher level of scriptural secrets that most people may not believe. This book draws the attention of the world toward the pseudepigrapha (a collection of hidden and rejected books, yet filled with deep secrets still valuable today) and explaining how, since thousands of years, God has already revealed deep details about the supernatural origin of the universe, but people (including those who believe or claim to believe in Him) have just refused to literally accept God's mysterious story of creation, which can never be understood by just sticking with conventional science. If you believe in God, have some origin-related questions whose answers you cannot find anywhere, not even in the Bible, and if you want to tap into historically neglected revelations to answer fundamental universe and life questions, then be sure to get a copy of *"Origin of the Spiritual World"* today. Learn more at Science180.com/pseudepigrapha

5. **"FROM SCIENCE TO BIBLE'S CONCLUSIONS"** (I called this book the "public version" of my book on the origin of the universe and it is tailored for the general public, and it is a great summary of the scientific version from a perspective that laypeople will fully understand). In this book, I, Nathanael-Israel Israel, broke down the complicated (scientific, philosophical including religious) data about the origin of the universe in a simple language that the general public can fully understand, and know in

order to live happily forever. Quickly grab and read this scientifically verifiable, bestselling book to finally get the accurate, jaw-dropping answer that has been rationally shaking believers, skeptics, and all freethinkers. Don't wait! Learn more at Science180.com/public

6. "*TURBULENT ORIGIN OF LIFE*" (This is the biological or life version of my book on the origin of the universe). It is meant to suit scientists, nonscientists, and all kids of laypeople, and it decodes the origin of all forms of life so human beings can understand and better live. As of 2025, this book is my only book devoted to the origin of all forms of life, and it will help you to grasp in a simple language what is needed to fully understand the formation of all forms of life. Whether you are a scientist or a layperson, a believer, or a skeptic, you cannot afford to ignore the greater, better, faster, simpler, cheaper, easier, and accurate formula unlocked in this important book that successfully decoded the origin of life. Get "*Turbulence Origin of Life*" today and change lives. Don't wait. Learn more at Science180.com/life

7. "*HOW BABY UNIVERSE WAS BORN*" (How was the universe formed? Did God really form it like some people believe, or did it come out of some long processes? How can we scientifically prove and break down this difficult mystery in a language that children will fully understand and like?) Get the answers as you read this book that I called the "children version" of my book on the origin of the universe and life. Accurately explaining the complex formation of the universe and of life to children can be very hard in our modern world, but by getting "*How Baby Universe was Born*," you will know the proven formula to help children to easily understand their huge universe-origin and life-origin questions with confidence, humor, and joy. They will surely belly laugh and thank you for it! It is time to buy this pragmatic book and offer it to the children in your life today. Learn more at Science180.com/children

8. "*HOW GOD CREATED BABY UNIVERSE*". The most difficult part of writing scientific things for children is how to break down complex technical concepts into simple words that they and even anyone who can read can clearly understand (without losing the accurate details and facts). When the topic to address is about the origin of the universe, the task is even more challenging for most people, but not for Nathanael-Israel Israel. As long as you can read, you will find this amazing book extremely helpful to grasp all complicated concepts needed to properly crack the origin of the universe in a language that even children ages 7-12 and anyone who did not go very far in school can fully comprehend.

Nathanael-Israel Israel: Has had the Honor to be Acknowledged the First Human Being that Scientifically Reconciled Science & the Biblical Creation

9. *"SCIENCE180 ACCURATE SCIENTIFIC PROOF OF GOD"* (Whether you are a believer, an unbeliever, a freethinker, an administrator, a politician, a curriculum designer, a curriculum specialist, an education policymaker, a librarian, a school board member, a parent, a researcher, a student, a teacher, clergy, or a layperson, as long as you are really seeking to scientifically understand the rational proof of the existence of God, *"Science180 Accurate Scientific Proof of God"* is the much-admired book written for great people just like you). As long as you are interested in the first and the only scientific book that talks to anti-creationists, evolutionists, big bang proponents, atheists, and all other freethinkers and rationalists about the universe's formation and they bigly beg to know more about God, the creator, whom they mistakenly deny; then this book is for you. As long as you are really seeking to scientifically understand the rational proof of the existence of God, *"Science180 Accurate Scientific Proof of God"* is the much-admired book written for great people just like you. Grab it today and start reading it. Don't wait any longer! Learn more at Science180.com/godproof

If you want to have the entire big picture of my discovery of the origin of the universe, life, and chemicals, and to enlighten your life and career, then plan to get all or some of these books that best suit your needs and interests. For more details, visit Science180.com/books

CHAPTER 15

HOW TO TALK TO EVOLUTIONISTS, BIG BANG PROPONENTS, ATHEISTS, AND ALL OTHER FREETHINKERS ABOUT THE FORMATION OF THE EARTH-MOON SYSTEM, AND THEY WILL BEG YOU TO TEACH THEM MORE ABOUT GOD, THE CREATOR

In addition to the general processes that I already demonstrated and explained about the formation of the celestial bodies in the universe, the Bible provides more information about the formation of the Earth, the Moon, and the Sun. In this chapter, I will provide additional details pertaining to the Earth-Moon system. With that being said, this chapter is the continuation of the previous one in which I explained how the scientific data perfectly agree with the Biblical timeline and story of the formation of the Earth (on the 3rd day), and the Moon and the Sun on the 4th day of creation. If you have not yet read the scientific demonstration of the birthdates of the Earth, the Moon, and the Sun, please read the previous chapter. This chapter will also help you understand how, by sticking to the word of God, I progressively unearthed the code for the formation of the universe long before I discovered the mathematical formula for the birthdates of the celestial bodies.

Like I explained, before forming the Earth, God laid its foundation and properly measured its precise dimensions: "*Where were you* [Job] *when I* [God] *laid the foundations of earth? Tell Me, if you have understanding. Who set its dimensions - if you know - or who stretched a line over it. On what were its foundations set or who laid its cornerstone*" (Job 38: 4-6). This statement implies that the dimensions of the Earth's precursor and, likely, those of other bodies were precisely set on specific foundations. The dimensions of the Earth's precursor must have had a significant impact on the Earth's size, movement, and other characteristics. Had the Earth's precursor been bigger or

Nathanael-Israel Israel: Known as the "Unquestionable Expert Witness for the Defense of Biblical Creationism" smaller than what it was, the status of Earth could have been different and the chemistry that supports life on Earth could have been different. In other words, the foundations that were laid for the Earth considered chemical parameters, which in the end defined the makeup of all of the particles and chemical elements on Earth. In my book on the origin of chemical particles, I showed how the chemical composition of celestial bodies depends on their position relative to the Sun, confirming that a different Earth position could have changed its makeup and the properties of life. Therefore, the dimensions of the fluid layers of the precursors of the Earth were not random but consistent with God's predetermined plan, according to which the universe could have a place for a planet called Earth, where living organisms, including human beings (which would be formed in a few days in the image of God) could live.

15.1. Tipping and tilting of the fluids of the Earth's precursor to form a solid body

The Bible suggested that, at one point during the formation of the Earth, God tilted (tipped over) the water jars of heaven so that dust (gas) could become a mass of mud and its clods stick together: *"Who can count the clouds by wisdom, or tip over the water jars of heaven, when dust hardens into a mass and clods of earth stick together"* (Job 38:37-38). I can only imagine the image Job had in mind when God was speaking to him. In those days, jars were how people carried water from wells, rivers, and lakes. In that time, the size of the Earth had never been seen from space. Job may have been looking around at what he could see and the areas he had been to, and thinking of what size water jar God was talking about. Looking at the pictures from space, one can only imagine how big those jars of water were.

On December 18, 2013, it appeared to me that this statement in Job 38:37-38 implies that, at one point, the precursor of the Earth was like a mass of water, whose tilting caused the waters to amass and remain together. Fluids such as water and gas could have been present at an early stage of the formation of the other celestial bodies. As I kept working on the origin of the universe, I later understood that as the fluid layers of the precursor of the Earth-Moon system moved, something caused them to gradually split, like a water jar when it is tipped over to overflow or change the movement of the fluids it contains.

On March 30, 2020, as I got a better understanding about turbulence, I realized that the above verse contains key information that needs to be carefully studied. Some of the main keywords in this statement are clouds, water jars, the tipping over of water jars, dust, the hardening of dust, mass, clods of earth, and the sticking together of clods of earth. Although no detail is given about the meaning of these keywords in the contemporary scientific language, some inferences can be drawn from them. In general, a clod, dust, or water mentioned in this statement is a kind of fluid, meaning that the statement addresses how the fluids of a body's precursor went through processes that changed its nature, including giving it mass and cohesion. Indeed, this quote implies that there were many water jars, which can mean that there were many bodies of water at one point. It can also mean that the verse is not addressing just the precursor of one body but of many bodies. The statement suggests that the tipping over of the water jars was associated with the hardening of some dust, which ultimately gained weight or mass. The tipping over of the water jars could also allude to the split-gathering and/or the events that tilted or inclined the precursors of the bodies, their orbits, or other characteristics related to their movement. Regardless of the meaning of these events, the statement suggests that the tipping over of the water jars and the hardening of the dust into mass and the sticking of the clods of earth were connected and/or are processes which could have occurred at the same time.

The same statement implies that as the dust was hardening, the clods of earth stuck together. This suggests that the precursors of the bodies stuck together after their fluid bodies were tipped over, and a process was initiated that caused their hardening, leading to the formation of other structures called clods, which ultimately stuck together. The hardening of dust into a massive body could be alluding to the creation or formation of mass or weight. The clods, the water, and the dust mentioned in the statement are not the same thing, implying that each of these processes could have forced the clods or water of the precursor to go through some changes until the final matter was formed. The sticking together of the clods is a kind of creation or formation of a cohesion between the particles in the clods, and the formation of bonds that have binding forces or binding power that attach the otherwise dispersed clods into more united bodies. It can also allude to a force that could have compressed the precursors of the fluids.

As some bodies or clusters of bodies were bound to force them to stick together, others could have been loosed. The gathering together and the stretching of the bodies or their clusters could have applied not only to small bodies but also to bigger ones. For instance, the binding and the loosening of constellations were also introduced in the same story (Job 38:31): "*Can you* [Job] *bind the chains of Pleiades or loosen the belt of Orion?*". In Amos 5:8, God is also mentioned as the maker of Pleiades and Orion. Pleiades is known as the constellation of the 7 sisters while Orion is the constellation whose belt consists of 3 stars. Pleiades and Orion are some of the constellations known since antiquity. Contemporary evidence suggests that constellations consist of many stars, meaning that the statement above alludes to clusters of stars bound by some chains, while, when others were forming, some belts were loosened. It can also be fair to deduce from this statement that belts of stars or belts of celestial bodies could have been caused by the loosening of something that had bound them together. In other words, the ability of stars and other bodies to cluster together could have been impacted by a force that bound them or loosened them otherwise.

15.2. Division or separation of the water layers of the precursor of the Earth

The Book of Genesis talks a lot about waters referred to as "Mayim" in Hebrew. However, it seemed to me that what Moses was referring to as waters in Genesis 1 could not have been just water consisting of hydrogen and oxygen atoms, but water mixed with other chemical elements and/or their precursors. Many other chemical elements could have been formed by the time the Earth was formed. Like I said earlier, in the days that Moses revealed the creation narrative, nothing was known about chemistry. Even more than 1,000 years after the writing of Genesis, early Greek scientists still classified everything in the universe using only water, fire, air, and soil. In other words, Moses could not have done a better to expressing the fluid-like precursor of the Earth.

According to the Bible, the waters that surrounded the precursor of the Earth were divided or separated into different compartments (Genesis 1:6-8) and then the sky was created in the opened space. During the formation of the sky, a portion or layer of water was moved above the sky while another portion was positioned below the sky. The Earth was formed using the waters that were below the sky. The story in Genesis 1 did not specifically say the outcome of the water which went above the sky. Other references suggest that the water above the sky could have been the water that rained on Earth during the days of Noah, when the flood occurred as God opened the window of heaven (Genesis 7:11).

The Bible does not say whether, during the great flood in the days of Noah, God poured out all the water that covered the sky onto the Earth. However, the flood story recounts that at the end of the 40th day of rain, God shut the window of heaven and the rain stopped. This implied that, during the early years after creation, layers of water could have surrounded the Earth like how the layers of ozone and other gases surround the Earth today. It is possible that the mass of water above the sky refers to the mass of fluids even beyond the Solar System, for the location where the Sun was formed (called Rakia in Hebrew) seems to denote the space between the waters above and the waters below. The narrative of Genesis 1 seems to emphasize the formation of the Earth. Hence, human beings have been having a hard time explaining Genesis 1 using the current scientific knowledge that has provided information far beyond the visible night sky.

In the days of Moses, because all that people could see were stars in the night sky, his story may be relating to the things visible in those days, which is significantly less than what telescopes see; it is possible that Moses focused on that limited sky. Today, we know that the stars visible to the naked eye are a tiny portion of the stars in the Milky Way galaxy. However, the creation story in the Book of Enoch (that I detailed in *"Origin of the Spiritual World"*) described eons or things that were precursors of the current universe in a language that even the most advanced contemporary scientists may not understand. For those revelations are about things that scientists will never physically see, for they no longer exist, and no physical terms or experience could allow scientists to properly describe them without relying on faith.

Finally, before I close this segment on the fluids of the precursor of the Earth, it is important to know that the waters mentioned by Moses were not brought to their place or position by any other celestial body (such as comets, which some people think are the source), but were formed by God. As God was separating the waters from the waters on Day 2, He could have ensured that the proper amount remained below the sky; otherwise, the size of the Earth could have been different, and many environmental factors could have varied in ways that would have affected life. A bigger or a smaller Earth could offer different conditions for life. In other words, God's design and leadership during the formation of the Earth ensured that everything was done to meet His plan. As these ideas came into my mind on December 25, 2013, I did not even know that a day would come when I would be writing about the origin of the universe. In all humility, I know that my historic discovery was not by chance but the product of years of obedience to God and His word without forgetting how my career was sacrificed. Still, I consider all that a gain!

On May 10, 2017, months before I discovered turbulence in the data I was analyzing, it appeared to me that the division and the aggregation of the waters mentioned in Genesis 1 must have been very fast and occurred simultaneously as other precursors in the universe were being shaped. In other words, I felt that just as the precursor of the Earth was undergoing changes, the precursors of other bodies in the universe might have been doing the same, according to laws I was struggling to demonstrate scientifically in those days. But after more than 12 years of seeking, asking, and knocking, …, I now understand those laws and the principles behind them properly. Part of my mission is to share my insight with the whole world, and you can join me by registering at www.Science180Academy.com.

I remembered that, in early 2017, I used to take water in a bowl and throw it into the air just to see which physical mechanism could illustrate or explain how it broke up. If you take water in a big bowl and start moving it, you will see the water start moving, and, depending on the force you use, it can break and even spill out of the bowl before you intentionally throw it in the air. As I was doing this experiment around mid-2017, I noticed that the volume of the initial water, which I threw into the air, was split into many smaller volumes of various sizes. Some blocks or drops of water are small, while others are big. In those days, although I did not know the scientific processes by which the waters were breaking apart, I was amazed at how the creation story narrated by Moses reflects realities according to which an initial amount of water highly moved by the Holy Spirit could have been able to break up and separate into different layers and create a space or sky in between them. Looking back now, I can see that I was searching for the secrets in the Biblical story of creation with the faith of a child, amazed at the little progress I was making as I pieced together the creation mysteries.

In those days, I already understood that the direction into which the waters of the precursors were moved could have impacted the direction or sense of the revolution and rotation of the Earth. The precursor of the Earth could have moved in the direction it was moved or launched in. Considering the Earth's orbital and rotational speeds, I suspected that the waters of the Earth's precursor could have moved very fast, which also aided in their rapid separation from other waters. Because the Spirit of God is said to have been moving at the surface of the waters already in the first moment of the first day, and because it was on the second day that the separation or breakup of the waters of the precursors of the Earth was reported, I deduced in those days that it could have taken 1–2 days maximum for the breakup to happen. Although these timings may sound trivial to some, they are key to unraveling the high-level mathematics encoded behind what Moses said thousands of years before science was even created.

The distance separating the fluid layers after their separation could have also depended on the speed with which they were separated. In other words, the vast distance separating the celestial bodies could also indicate the high speed with which their precursors were separated. On the scale of the Earth, as the layers of water were separating, different amounts of water were being organized into larger amounts, leading to the formation of various chemical compounds embedded in minerals, rocks, and mineraloids.

In the days when I was first inspired to illustrate fluid separation using moving water in a bowl, I marveled at the sight of water breaking in the air after I threw it. That was an "aha moment" for me and no one around me could fathom why I was so thrilled at trying to understand a simple creation story that billions of human beings before us have read. Around mid-2017, I tried very hard to figure out the scientific phenomenon that can explain water separation. It took me months of reading before I realized it was related to properties of fluid dynamics. I had to read hundreds of articles about fluid splashing, sloshing, and fluid breakup before I realized that a phenomenon called "turbulence" could have been in play. It took me almost 3 years to finally understand what could have happened. More details about that scientific journey can be found in my forthcoming autobiography, and also in *"Turbulent Origin of the Universe,"* and *"From Science to Bible's Conclusions."*

In my book titled *"Origin of the Spiritual World"*, I explained how the lost scriptures (such as the Book of Enoch) also revealed that God formed the Earth by piling up rocks. That story also suggests when rocks were formed already. I have to refer to the Book of Enoch here because, besides that book, no other scripture clearly explains how God piled up rocks to form the Earth. Hence when we dig the land anywhere on Earth, rocks are found below. In some countries, roads are built by blasting rocks with dynamite. The Earth is indeed built on a firm foundation. This information about the piling of rock to form the Earth is very consequential, for it also implies that the formation of chemical elements could have been completed before the completion of the Earth's formation. Furthermore, if on the 3rd day of creation, plants were already growing and reproducing, it also implied that all the chemical elements found in plants had already formed. The setting of my mind on these facts allowed me to reject theories that alleged the formation of chemicals as a matter of millions of years or as a matter of a migration of compounds from other celestial bodies. By the 3rd day of creation in which the formation of the earth was finished, the Sun that some false theories claim to be the source of most of the chemical elements was not even fully formed yet.

15.3. Formation of the Earth's crust and mountains

During a Bible study on February 19, 2015, that I did by myself, I was inspired by the idea of how mountains could have formed. According to the Bible, the formation of the Earth was completed on the 3rd day with the removal of waters covering the face of the Earth and their gathering together to form the seas so that the dry land could appear (Genesis 1:9). Knowing that on the second day of creation the Bible also talks about the separation of waters from waters and the formation of the sky, I understood that the Earth mentioned in Genesis 1:1 must have been a precursor of the Earth. Else, additional features indispensable to the Earth, such as the continents, oceans, and sky, could not have formed after Day 1. I also deduced that the formation of many other celestial bodies could have taken some days. I was not surprised to realize that the formation of the Sun and the Moon was completed the next day after the Earth's formation. The formation of the dry land on the 3rd day made me think that some of the mountains on Earth could have formed by then. In other words, the final shape of the Earth was acquired by the end of the 3rd day.

For the waters to recede from the continent so the dry land could appear, God could have broken part of the crust so the water could flow inside the opening. While water started entering the ground, the latter could have been pushed toward the continent, birthing mountains near the coasts. In other words, as some mountains were forming, the waters covering the Earth were moved inside the holes of the oceans. As more water entered the oceans, the oceans could have been pushing the continents and distancing the lands that are near the openings from one another. In other words, the elevation of the mountains and the birth of the valleys and depth of the oceans could have occurred simultaneously as insinuated in Psalms 104:6-9.

The process I just demonstrated could have also occurred during the great flood in the days of Noah. Indeed, during the flood of Noah, the heights of some mountains may have increased, allowing excess water to find a place and dry land to appear. In those days, the oceans gained more water, and their depth could have increased. In other words, the process by which the flood of Noah receded from the continent could have had some similarities with how the mass of water covering the Earth in Genesis 1 was gathered together so the dry land could appear. New mountains could have been formed for the flood of Noah to recede and existing mountains could have gained more height. The weight of the flood of Noah could have been heavy on the Earth's crust and could have cracked some crusts and sent some water beneath the Earth. If God had not properly calculated the Earth's foundation, it could have broken down either during the Earth's formation or during the flood of Noah. An incorrect design could also have exposed the Earth's plates to greater tectonic activity and endangered life on Earth. Glory be to God for His design!

On November 25, 2013, I felt that the wind God used to dry the Earth on the day of the flood of Noah (Genesis 8:1) could also have been used during creation in Genesis 1. That wind could have contributed to breaking the Earth's crust at some locations, while pushing some waters all the way to the poles where they were stacked into icebergs, which recently have been melting at an alarming rate that is defying scientific modeling, while the water from the rising oceans is causing more flooding of beaches and countries are losing coastlands as if the reverse of what God did to form the continents and oceans is happening in our time. Yet people are not paying attention! When natural catastrophes (e.g., earthquakes, floods) that Jesus predicted would happen before the end are occurring in our day, people do not pay attention to the fact that the return of the Maker of the heavens and earth, who knew what He had created, is near!

On December 18, 2013, I also realized that clouds are in the sky for a reason including the preservation of heat on Earth. Indeed, that day, I learned that God made the clouds as a blanket or garment for the sea (Job 38:9). This means that clouds prevent the sea from drying because of the heat of the sun or from getting too cold and freezing. A blanket or a garment is used to protect something. Clouds also protect the Earth just as how a blanket protects a human body and keeps it warm. A blanket prevents humans from getting cold. Without clouds, the Earth would be colder. Heat may be reflected, dissipating into the atmosphere, while clouds keep it within Earth's atmosphere. The cold is trying to come from space and enter Earth's atmosphere (e.g., in wintertime), but clouds help retain heat on Earth. The atmosphere of the Earth must be special. Fog on Earth is also a type of cloth. On nights when the Sun is down and the Earth starts cooling, darkness also helps to keep the Earth cooler until another day comes. God commands the day and taught all seasons to know their time (Job 38:11-13). If God did not design a mechanism so that evaporating water does not leave the Earth indefinitely, but instead precipitates and falls back on Earth, we could have lost most of the Earth's water by now, causing a catastrophe for the Earth and the life it hosts! Did you think about that and thank God for that?

Among all of the celestial bodies in the universe, none other than the Earth was specified as God's footstool, pointing to the importance of the Earth in the divine planning: *"Thus saith the Lord, the heaven is my throne, and the earth is my footstool: where is the house that ye build unto me? and where is the place of my rest?"* (Isaiah 66:1). All of the earth belongs to God for a reason:

- *"Now therefore, if ye will obey my voice indeed, and keep my covenant, then ye shall be a peculiar treasure unto me above all people: for all the earth is mine"* (Exodus 19:5).

- *"The heaven, even the heavens, are the Lord's: but the earth hath he given to the children of men"* (Psalm 115:16).

The Earth is special, and we are special. Therefore, those who deny the existence of God are like tenants of a beautiful house who reject the existence of their landlord (who lives in a remote place that they ignore or that they cannot reach). Because God revealed Himself in the Bible and in nature (Romans 1:18-32), unbelievers are like tenants who refuse to follow the guidelines and instructions of their landlord despite being aware of the manuals left behind by the owner of the property that they think came into existence by chance or through billions of years of process. For those who believe, we recognize that God is our Maker and He deserves to be praised and believed even when we don't always understand everything in nature!

15.4. Formation of the Moon according to the Bible

In addition to the general process that explains the formation of the satellites, this segment expounds on the formation of the Moon using Biblical information. Indeed, a lot of money and effort have been invested into speculating on the origin of the Moon, and a lot of debates and theories have tried to solve this mystery. However, before the publication of my findings, no one had comprehensively explained how the Earth-Moon system was formed.

A few man-made missions have been deployed to the Moon, and although the chemical composition of some lunar rocks has been analyzed, many questions still remain unanswered regarding the Moon's origin. Unfortunately, most investigations focused either on the Moon only, or the Earth only, or the Earth-Moon system only, without integrating their origin with that of the other bodies in the Solar System, which cannot be comprehensively understood if its elements are separated and handled in disparate fields of research whose authors don't even communicate much. That was one of the challenges I tackled in my approach.

According to the Bible, the formation of the Moon was completed the next day after the completion of the formation of the Earth. This implies that when the Earth was forming, the Moon's precursor may have been undergoing its own formation. In other words, although the formation of the Moon was completed on the 4th day of creation, it is more likely that by the second day of creation, when the waters were splitting from waters, the precursor of the Moon was also going through the process of its formation. This is like a woman's pregnancy: when a baby is conceived, he or she begins to take shape, and after months of development, the baby finally reaches its final form. Considering what I know now, I think that the split of waters from waters mentioned in Genesis 1 also alludes to the split of the precursor of the Earth-Moon system from the rest of the precursors of the bodies orbiting the Sun, and also the split of the precursor of the Moon from the precursor of the Earth, without forgetting the split-gathering of the precursor of the Sun. In fact, before the second day of creation, the precursor of the Earth-Moon system was sandwiched between the precursors of bodies above it and those of the bodies beneath it.

Although the Bible did not describe these processes, the stages through which the precursor of the Earth passed before yielding the Earth indicated how the Moon could have been formed. Long before I discovered turbulence in the scientific data, I knew that just as the Earth was not formed instantly but over 3 days, so also the Moon was not formed suddenly but over 4 days. On September 26, 2016, I first wondered whether the mechanism that gave rise to the Earth and the Moon could be used to explain the formation of all planetary systems (systems consisting of a primary planet and its satellites). In those days, I neither had nor knew of any logical explanation for the scientific process by which the Earth-Moon system formed. But about five years later, as I got a better understanding of the formation of the universe, I solved that problem.

Although the Bible did not expressly say it, the waters that were moved above the sky by the second day of creation may have included the precursor of the Moon. The formation of the Moon was completed on the 4th day because it took time for the precursor of the Moon to travel from its position in the precursor Earth-Moon system to its current position, while the precursor of the Earth was undergoing its changes. Putting together the Biblical story of creation and the scientific data, I concluded (long before I discovered the formula of the birthdate of the celestial bodies) that the Earth and the Moon descended from the same precursor, which, after splitting into the precursor of the Earth and the precursor of the Moon, disappeared, and let each of these precursors to "mature" and yield each of these celestial bodies, the Earth by the 3rd day and the Moon by the 4th day.

Unlike the Earth, the Moon was not made to be inhabited but to serve as a measuring stick for signs and seasons and to dominate the Earth at night (Genesis 1). For instance, the Jewish cultural calendar is based on the Moon and the Sun. Many Jewish feasts or festivities are based on the Moon. Hence, the phases of the Moon, particularly the new Moon, have been used in Jewish rituals, including the beginning of some feasts, the time to offer certain sacrifices, the blowing of the trumpet (shofar), etc. Although as of today, the Moon is indispensable for life on Earth, providing light at night, it is important to note that the New Jerusalem to come down on Earth will have *"no need of the Sun, neither of the Moon, to shine in it: for the glory of God will lighten it, and the Lamb is the light thereof"* (Revelation 21:23). What a great and glorious time that will be when the Lamb of God will be the light of the day and the night! There will be no more darkness. God is the *lamp of the believers, and He turns their darkness into light* (2 Samuel 22:29). John, who wrote several books in the Bible (the gospel of John and the 3 epistles of John as well as the book of Revelation) and who was a disciple of Jesus, declared the following:

> 1 John 1:5 *This is the message we have heard from him and declare to you: God is light; in him there is no darkness at all. 6 If we claim to have fellowship with him yet walk in*

the darkness, we lie and do not live by the truth. 7 But if we walk in the light, as he is in the light, we have fellowship with one another, and the blood of Jesus, his Son, purifies us from all sin. 8 If we claim to be without sin, we deceive ourselves and the truth is not in us. 9 If we confess our sins, he is faithful and just and will forgive us our sins and purify us from all unrighteousness."

To see that day, we must believe in God and surrender our lives to walking in His light. When we do that, we will see in the end what Jesus (Yeshua) was talking about in John 8:12 as he was speaking to the people: "*I am the light of the world. Whoever follows me will never walk in darkness but will have the light of life.*" One day, we, the believers in Jesus, will have the light of life he was talking about, and He will be our light both in the day and in the night and forevermore! Until then, remember that God created the Earth and the Moon on the 3rd and 4th day, respectively. In the chapter on the formation of light, I will revisit this concept.

'Science180 Academy' Success Strategy
SCIENCE180 SEMINARS

People whose awareness is raised by Science180 usually ask me to go deeper or they wonder "what else?". That is one of the reasons Science180 trains them through strategic work sessions (during seminars or training sessions) that transfer customizable skills and solutions to them. Science180 Seminars are client-centered and tailored to strongly engage clients, helping them maximize discovery and tap into new opportunities to outperform their expectations. Science180 offers customizable seminars that can be labeled as a colloquy, conference, consultation, discussion, forum, keynote speech, lecture, lesson, meeting, symposium, summit, study group, tutorial, workshop or working section accordingly on any topic related to:

- Universe-origin for scientists and mathematicians, philosophers, laypeople, and the general public
- Universe-origin or universe creation for believers
- The origin of life for life scientists, for all other scientists, and for believers
- Chemical-origin for scientists
- Universe-origin seminars for children
- Universe and life-origin for pseudepigraphic believers

As you contact us with your needs, we can customize your program accordingly. Learn more at Science180Seminars.com.

CHAPTER 16

HOW SMART PEOPLE SCIENTIFICALLY DECRYPT THE PROOF OF THE MYSTERIOUS MISSION OF THE SUN

As I carefully studied the Sun, I realized that its formation and motion hide an encrypted code connected to the Earth and life. Indeed, in previous chapters, I discussed the processes that the precursors of the celestial bodies, including stars such as the Sun, underwent during the formation of the universe. I also scientifically showed how the Sun was formed on the 4th day of creation. If you need to refresh your mind on the scientific proof of the Sun's formation on the 4th day, revisit the chapter on the duration of the formation of celestial bodies. Using the biblical account of creation and the scientific data, I mainly tackled in this chapter some specific mysteries relative to the Sun and its domination of many things on the Earth.

Although the Sun is the brightest object that human beings can see in the sky, many questions about its origin still remain. Since antiquity, human beings have been seeking to understand the origin of the Sun. Because of the mysteries surrounding its origin, many people have thought that the Sun is God; therefore, they have been worshipping it for millennia. On almost all continents, people have idolized the Sun. On the contemporary or secular calendar (i.e., the Gregorian calendar), the first day of the week is even dedicated to the Sun: Sunday. Even in western countries where many people seem to deny the spiritual world (yet they practice occultism "without knowing"), archeological data show that dead civilizations which lived in the US devoted a lot of attention to the Sun. For instance, archeological vestiges at the Cahokia Mounds site in Illinois, near Saint Louis (the largest city in the state of Missouri), showed that the civilization which lived there worshipped the Sun. Archaeologists believe the Cahokia Woodhenge (which is large wooden poles) was used for agricultural reasons to tell the seasons as well as for religious observances. I was astounded when I visited that site in 2013. Other groups outside the US, like the Aztec Indians, were also known to worship the Sun, the Moon, and Venus. The Sun is also used to tell time; on ships, people used a sundial.

As of today, many countries have been spending billions of dollars to understand the Sun's origin and related phenomena, yet they have not succeeded. Unfortunately, because of its heat, the Sun is one of the bodies in the Solar System which surface has not been visited by human beings or man-made spacecraft. Many spectroscopic studies and flybys have been performed near the Sun, and in 2021 as I was editing this chapter, some advanced studies were being conducted on the Sun. However, a lot still needs to be explained. Many theories have been proposed to explain its origin, but they have failed.

The Bible states that the Sun was formed on the 4th day of creation, but my comprehensive review of Genesis 1 suggests that its formation began on the 1st day of creation. The Hebrew word "kokhavim" or "*kokabim*" used for "stars" in Genesis 1 refers to any bright object in the sky, including other stars, planets, asteroids, meteors, etc. Unlike the Earth, whose origin was more detailed in Genesis 1, a lot is not said about the formation of the Sun. The Bible underlined the Sun's mission to dominate the day and serve as a sign for the seasons, not just seasons defined by weather forecasts, but also prophetic seasons defined by appointed days and feasts.

I could not talk about the codes I decrypted about the formation of the universe without briefly mentioning here some of the early thoughts that launched me into this investigative journey. For instance, at 2:30 AM on November 9, 2013, I lost sleep, and I found my mind pondering so many questions that some people may consider trivial:

- What would have happened if the night had never existed?
- Why does the night fall?
- Why does the Earth rotate?
- Why does the Sun radiate light at its surface, while the Earth does not?
- Is the rotation of the Earth connected to the Sun?

These were the first questions I ever asked concerning the movement of the planets and the stars. Without any proof to back it up, I sensed in those days that the understanding of these questions could lead to a major discovery. As ideas were crossing my mind, I woke up from the bed to go record them in the living room. That night, I understood that if the night did not fall until about 6 PM and stay until about 6 AM, life could have been different from what it is now. The night also allows the Earth to cool down. During the day, the Earth's heat increases due to solar radiation, and the temperature of most things that receive light from the Sun can rise. And without the night, these objects on Earth would be heated and even overheated. Some chemical elements could have even already received enough light to melt them. Unless living things existed that could respond and adjust to the increase of the heat associated with days without night, some nonliving things could have disappeared. The incessant increase in heat and Solar energy could have also affected many geological, chemical, and biological cycles. To make a long story short, the succession of days and nights is very important for the maintenance of life on Earth. And if the processes that formed the Solar System did not allow the rotation of the celestial bodies, but just a linear movement, life would not have been the same. If the Earth were not rotating on its axis every 24 hours but were stationary yet maintaining its revolution around the Sun, the side of the Earth facing the Sun might be so hot that living things could not survive very long, as they would have no time to cool down. At the same time, those who are on the opposite side of the Earth may be so cold and never have the time to warm up. These factors would also affect the needs of the human body, such as vitamin D and sleep, without which human beings would suffer from depression, insomnia, and other health conditions. Hence, I know that the rotation of celestial bodies at specific speeds was also planned by God.

As of today, the North and South Poles have almost 6 months of night and 6 months of day, which, I guess, does not make life that much enjoyable for those who live in those areas, although by now they are used to it. If the size and rotational speed of the Sun were different, the length of the day and night could also have changed. Consequently, human beings would not have been what they are today, for they would have been forced to deal with longer or shorter days and nights than their physiology could allow. Although in those days of 2013, I never thought about writing a book on the formation of the universe, I was led to think about the following:

- the equilibrium between the length of the day and human activities
- the impact of the rotation and revolution of the Earth on the control of the energy it receives from the Sun
- the encrypted code of how the rest of the people get in the night synchronizes human activities with natural laws
- how fatigue and rest hide a code associated with the Sun, the Earth, and life

Indeed, many living things on Earth depend on the Sun for their needs. For instance, without the Sun, most plants could not obtain the energy needed to produce the biomass that most animals depend on. Considering the importance of plants for humans, I understood that, without the Sun, life on Earth, at least for human beings, would have been impossible. This implies that God had calibrated the size, position, composition, radiation, quality, and quantity of light, and other characteristics of the Sun so they match the needs of things it would dominate. I also came to realize that, if the size of the precursor of the Solar System was different, the bodies in the Solar System could have had a different size, therefore a different Solar System would have been formed. Consequently, the Earth and its inhabitants would have been different. In other words, the precision of the size and speed of the precursor of the Solar System was very accurate so that human beings could live on Earth. God may not have devoted the same care to the formation of all other stars. I understood with my studies that the length of the day and night are in harmony with life and natural laws.

As I thought about these questions, I realized that the rotation and revolution of the Earth help control the solar radiation received by the Earth. On November 10, 2013, meaning the next day after I first had the aforementioned racing thoughts, it appeared to me that the amount of energy received by the Earth depends on the seasons and its position in space. In general, the energy received in the morning and evening is inferior to that received around noon. The side of the Earth that is in the night cools down, while the side in the day warms up. To cool down, the Earth also radiates part of the energy it had received from the Sun back into the atmosphere. If the Earth's materials were not designed to allow such reflection, the Earth's temperature could have been increasing incessantly, causing problems for living things. Likewise, the nature of the atmospheric particles allows for the transfer of energy between the Sun, the Earth, and the Earth's atmosphere, etc.

Although greenhouse gases in the atmosphere are blamed for contributing to global warming, other theories also suggest global cooling, and the changes in Earth's environmental conditions are not yet sufficient to stop or halt life. If the surface of the Earth where human beings live were not shielded from the hot lava or magma in the Earth's interior, heat from the interior could already have decimated living things. In other words, the design of the Earth's crust and other internal layers agree with life just as the design and positioning of the Sun also allow the right amount of radiation to reach the Earth. Otherwise, human beings could have been sandwiched between the scorching heat of the Sun and the heat of the interior of the Earth. The Earth's crust's hardness, which prevents it from collapsing, is another key element God considered in His design. For this reason, allow me to pause for a second to thank God, on behalf of all His creation, for his masterpiece and awesome design! Thank you, God!

Just as the rotation of the Earth allows daily adjustments to environmental conditions (e.g., temperature), so also do the seasons encoded in the Earth's revolution around the Sun contribute to adjusting the environmental conditions and needs of living things from one year to the next. For instance, the alternation of hot seasons with cold seasons is indispensable to balance the impact of cold and heat on Earth. For instance, on its journey toward its aphelia (farthest point from the Sun), the Earth experiences colder days, while on its journey toward its perihelion (closest point to the Sun), the Earth gets hotter. The Earth's motion around the Sun helps balance environmental conditions. In other words, the "extra" heat received by the Earth while in proximity to the Sun is reduced by the "extra" cold received far away from the Sun. At the same time, the "extra" heat received when a pole is facing the Sun can be reduced by the "extra" cold it receives away from the Sun. In other words, winter reduces the impact of the heat of summer and vice versa, while spring and fall are transitions between the extreme heat and cold. The fact that, in general, when the Northern Hemisphere is experiencing its hottest season, the Southern Hemisphere is undergoing its coldest season is not by chance but was to ensure that the Earth is still habitable despite the variation of the environmental conditions.

As I said earlier, if the Earth were not rotating, one side could have been frozen or remained very cold, while the other side facing the Sun could have been burning with heat, making it hard or impossible for human beings to live on both sides. Likewise, if the Earth were moving on a linear trajectory away from the Sun, its temperature could also have reached colder levels incompatible with life. If the Earth were moving only toward the Sun, its temperature could also have reached levels hotter than those detrimental to life.

I later understood that the rotation and the revolution are connected and together they hold an encrypted message concerning the maintenance of life on Earth. For without them, life could have been impossible and/or human beings could not have kept their current characteristics and still live on such an alternative Earth that would rotate or orbit differently. To make a long story short, the movement of the Earth and the Sun allows environmental conditions to change, giving living beings diverse climatic conditions indispensable for them to avoid living in a boring, monotonous world. All of this was God's design during His laying of the foundation of the universe. Yet, we take them for granted! For that ingratitude, I must apologize!

The foundation that God laid for the Sun and the Earth was also aligned with the rest and sleep that all human beings dearly need. Indeed, to function well, all living things need to rest periodically. Medical studies show that human beings need at least 8 hours of sleep every day, according to their age, profession, and many other characteristics. Children sleep more than adults. Because they are generally quieter than days, most nights prompt human beings and other living things to slow their activities so they can rest more. Exceptions can be made for some big cats and other nocturnal animals that are active at night, while most herbivores (their food) are sleeping. Even in Western nations where many people work evening and overnight shifts, traffic is usually heavier in the day than at night. In general, when one side of the Earth is sleeping, the other side is working. For instance, when most Chinese and Nepalese are sleeping, most Americans are working, and vice versa, for their nations are geographically on opposite sides of the Earth.

Living things which are not human are also indirectly led to "rest" in the night. Plant activities in the night are different than those during the day. Sleep and rest were created to help living things normalize their behaviors and activities in accordance with natural laws, and no living thing on Earth is designed to function without periodically resting. Fatigue signals to living organisms that they need to replenish their energy or state of mind so they do not break down. In other words, rest and sleep are spiritual operators that trigger the renewal of energy needed to accomplish vital functions. Hence, those who do not sleep or rest can get sick and even mentally sick until losing their mind and life. Yet, God Himself never slumbers nor sleeps (Ps 121:1-8), for He cannot get tired, or be out of energy, or need to replenish it.

If the design of the Sun and Earth were different, human beings and other living things could have been different. And because human beings were made in the image and likeness of God, the design of the Sun and Earth hides a divine, encrypted code about God Himself! In 2013, as I thought through the relationship between the Sun, the Earth, and the living things, I realized that, among many things, the melting point of chemical elements, the amount of solar radiation received by the Earth, the mass of the Earth, the distance separating the Earth from the Sun, and the ability of the Earth to absorb, transfer, and reflect some of the Solar radiation and many other characteristics of the Earth and Sun were calibrated accordingly and God deserves the glory, the honor, the power, and the praise for all of these.

Looking back at my journey, I am not surprised that the mysteries about the formation of the universe were progressively revealed to me as I was devoting myself to God and seeking His help in the midst of the problems I went through in my life and particularly from 2011 to 2013!

It is not by chance that God linked the blessing of his children to the probability of having day, night, winter, and summer as long as the Earth exists (Genesis 8:21-22). It is also not by chance that the Bible suggests a link between the Sun, the Earth, the Moon, and the stars, saying:

- *"But immediately after the trouble of those days, the Sun will be darkened, and the Moon will not give its light, and the stars will fall from heavens and the powers of the heavens will be shaken"* (Matthew 24:29).

- *"I saw when the Lamb opened the sixth seal, and there was a great earthquake. The Sun became as black as sackcloth made of goat's hair, and the full moon became like blood. The stars of heaven fell to the earth like a fig tree drops unripe figs when shaken by a great wind. The heaven ripped apart like a scroll being rolled up, and every mountain and island was moved from their places"* (Revelation 6: 12-14)

Although many Christians can point to many of the things I mentioned in this chapter, none of them has been able to scientifically explain how God created the universe. One thing is to talk about intelligent design, but another is to properly explain how that design and its implementation came about. Being unable to explain God's creation, many believers attack the unbelievers who are just seeking the truth they could not find anywhere else. At the same time, many believers are lying about things and processes that never occurred during the formation of the universe. I thank the Lord for the grace He gave me to tap into some creation mysteries, including scientifically demonstrating how the formation of the Sun was completed on the 4th day of creation (see previous chapters and more about the formation of the Sun in *"Turbulent Origin of the Universe"*). I also pray that God helps me stay in His word forever and ever.

'Science180 Academy' Success Strategy
SCIENCE180 CONSULTING

Because Science180's trainings, seminars, or strategic work sessions (through which it transfers skills and training solutions) are great, some customers want to go even deeper on a long-term, sustainable basis. That is where Science180 Consulting, one-on-one consulting, and mentoring (that some people may prefer calling coaching programs) comes in. That is where Science180 can truly change people's behavior on a long-term basis according to their specific needs. With Science180 Consulting, you will discover and understand the deep secrets of the formation of the universe, life, and chemicals around you. Hear Dr. Nathanael-Israel Israel's personal selection and teaching on key topics that will help you break the code of the universe's formation and functioning. All strategically designed to enlighten you, guide you to navigate and filter the massive data collected on the universe and its content so you know how to answer the world's most challenging origin questions, remove any scientific and philosophical cataracts that may be blocking you, and help bring you many steps closer to your best life today and forever. Science180 Consulting will train you, transfer unconventional skills to you, and change your behavior so you go deeper. To get started today or to learn more, go to Science180Consulting.com.

CHAPTER 17

CAN WE DEMONSTRATE WITHOUT A DOUBT WHY AND HOW GOD CREATED DARKNESS AND FORMED LIGHT?

17.1. Creation and role of darkness

Although many people consider darkness as a state of emptiness or nothingness, it is important to know that the biblical evidence supports that God created darkness, even before light. Even today, the Jews start counting the beginning of the day with the night or at sunset. Therefore, it will be unjust to talk about creation without properly addressing darkness. Most people have probably never really thought about darkness being created, but if there was not always light, then darkness had to be created. Just because there is darkness doesn't mean it is empty. The hallway in your house might be dark at night, yet it does not mean it is empty. In fact, there might be something in the hallway that even your smallest toe will find when you are trying to find your way to the bathroom down the hall in the middle of the night.

There are 2 types of darkness: physical and spiritual. In this chapter, my focus is on physical darkness: the darkness that can be seen by human beings and that contains nothing invisible belonging to the spiritual world. In other words, I mean a physical darkness denuded of everything. For like I explained later, as of today, many things are hidden in darkness. In the next chapter, I elaborated on physical darkness and spiritual darkness.

The physical darkness is one of the very first things that God created, and that was not fashioned into anything else. According to the creative narrative in Genesis 1, darkness was present at least "on the surface of the deep" before light was formed (Genesis 1:3): "*And the earth was without form, and void; and darkness was upon the face of the deep ...*" (Genesis 1:2). In other words, God created darkness before light, and that can also explain why, until today, the Jewish day starts with the evening, meaning the darkness. Some people may think that God cannot create darkness, for they associate darkness with evil things, yet God Himself said that He can create evil: *"I form the light, and create darkness: I make peace, and create evil: I the Lord do all these things" (Isaiah 45:7).*

A message encrypted in the above verse is that darkness was created, but light was formed, meaning something originally created by God had to be transformed or molded into something else to yield light. I will elaborate on that later. In other words, light is more complex than darkness, which bears witness to some of the things God created out of nothing.

After God created light on the first day, He separated it from the darkness: *"And God saw the light, that it was good; and God divided the light from the darkness"* (Genesis 1:4). Some renditions of that verse say that God *"distinguished the light from the darkness."* Others said that *"God distinguished light from dark."* The Hebrew word used to name the darkness is "choshech," while light was referred to as "ohr." In his account of the beginning, Apostle John also used the same term to refer to the light in Jesus, or the light of Jesus, as the light of men (John 1:4). Apostle John also used the same word "choshech" to refer to the obscurity or darkness that was unable to comprehend, overpower, put out, or grasp the light in Jesus. In other words, although seeming to explain different aspects of the creation of darkness and light from 2 different perspectives, Apostle John and Moses used the exact same terms to address light and darkness. Because people were living before Jesus was born, and those who rejected Him were already disbelieving beforehand, the account of Apostle John also confirms the fact that darkness existed before light.

Based on some data I calculated on the time it took to form some celestial bodies, I cannot argue with those who allude to or think that on the scale of the Earth, the separation between light and darkness mentioned in the first day of creation encodes the beginning of the rotation of the Earth and of other celestial bodies that have light on one side and darkness on the opposite side. Hence, God called the darkness night and the light day: *"And God called the light Day, and the darkness he called Night. And the evening and the morning were the first day"* (Genesis 1:5). This verse also backs the Jewish culture, which teaches that the day begins in the evening or at sunset.

Darkness has different intensities or thicknesses depending on the amount of light in it, if any. All nights are not the same. Some nights are darker than others. Biblical stories teach that some darkness can be thick enough to be felt physically, like the one that hit Egypt when the Pharaoh refused to let the Israelites go: *"And the Lord said unto Moses, Stretch out thine hand toward heaven, that there may be darkness over the land of Egypt, even darkness which may be felt"* (Exodus 10:21). Yet at the same time, there was light in the land of Goshen, where the Israelites were dwelling.

Likewise, as of today, some people are walking in darkness, while others in the same location are in the light. This fact also suggests that darkness is not just the absence of light. Darkness is a substance that can be called into existence even in the presence of light in such a way that, in the same place, some people may be in darkness and others in light. Likewise, depending on the location of the celestial bodies with respect to a primary star, the darkness around them at night can be different. According to other Jewish scriptures, the darkness in the second heaven is thicker than that on Earth today.

Because God is light and never changes, He could not have been in the darkness mentioned in Genesis 1. In other words, even before God formed light, the darkness that He created could not have been able to overpower or put Him out. Just as physical darkness always disappears when light shines, so also when light was formed, darkness disappeared and could be found only on the side of the bodies that were not receiving the light.

God's goal was not and is not to completely remove darkness from the universe. Otherwise, He could have designed a way for light to be seen everywhere, even all around the bodies that rotate, so that the night could not exist but disappear once and for all. Likewise, He could have also given a certain power to light to never leave room for darkness. However, God's plan was not to eradicate darkness from the face of the universe after He formed light.

Removing darkness could have even complicated some processes indispensable for the formation and maintenance of the universe. Just think if there was light all the time, some plants that need darkness to grow or perform some vital activities (e.g. respiration) would have a problem. In the same manner, as of today, God's goal is not to eradicate evil from the face of the world, but just as darkness ruled for a moment before light was formed, the day is coming when God will eternally separate spiritual people who bear light from those who have no light.

Just as darkness flees from light, so also some people flee from the truth that can save them. In other words, when light was formed, darkness fled away to hide itself where it could not be reached. Just as the created light did not chase darkness to destroy it, so also God, since the creation of the world, never intended to chase people to force them to change or believe in Him. This can also be why God chose to live in heaven and not among men so He does not force human beings to be in His light all the time or destroy some people before they can even be given a chance to live. Hence, trying to force everybody to behave the same way or in a certain way is like chasing the wind! This is a mystery that cannot be explained using mere physical means.

Hence, the characteristics of light do not allow it to adjust its behavior to the response the surface it hits gives. For instance, while light can penetrate certain things, it can also be rejected or deflected by some bodies it shines on. If the size, rotation, and revolution of the Earth were different, the length of the night or day could have been different, and in some conditions, night may not even exist. In other words, some characteristics of the Earth are also aligned with God's desire to allow darkness and light to exist on Earth, so that human beings who live there can also have a choice to do certain things in the night willingly or by "force", so that earthly conditions do not suppress some requirements for light. These encrypted requirements for day and night on Earth are also connected to some characteristics of the Sun, the Moon, and other stars. In the chapters on the formation of light and the Sun, I expounded on other aspects of darkness. In short, remember that light was formed, but the darkness was created. More details on the formation of light can be found in my book on the origin of chemical particles.

17.2. Formation and mystery of light

I could have addressed the formation of light long ago, but I felt that understanding its physical composition requires a little bit of background on the events that defined the turbulence, splitting, and the formation of fluid layers of the precursors of bodies and particles, including photons, the particles of light. Hence, I had to wait until this point before handling this crucial chapter filled with mysteries.

As I studied things in nature, I understood that there are 2 types of light: spiritual light and physical light. In the following lines, I explained these concepts and how they apply to the understanding of the formation and functioning of the universe.

17.2.1. Formation of physical light

The physical light was one of the things that God created on the first day of creation. Before the creation of the physical light, a spiritual light found in God already existed. This means that the light God created on the first day of creation was different from God's light (I will deal with this in another chapter very soon). Because God cannot create Himself, the light He created in Genesis 1 cannot be His own light. Likewise, the light of the world mentioned in John 1 as Jesus should not be confused with the light that God created in Gen 1:3. Therefore, those who worship the Sun and the stars are guilty of idolatry, which is ignorance (Isaiah 45:20).

The physical light that mere human beings can see was formed by God. In contrast to darkness, which was created, light was not created, but was formed, meaning something initially created was molded or shaped for light to appear. God Himself clearly made the difference between the formation of light and the creation of darkness when he spoke to Prophet Isaiah in these terms: *I form the light, and create darkness: I make peace, and create evil: I the Lord do all these things (Isaiah 45:7)*.

Light is known to consist of photons, which is defined as the particles of light. The formation of light implies that its precursor was shaped to form photons. Because light can be in different forms according to the nature, speed, visibility, and radiation, all types of light might not have been formed at the same time.

The visible light seems to be the focus of the narrative in the first day of creation. In other words, as the original matter that was the precursor of all matters was undergoing changes to yield the various types of matter in the universe, some radiations related to light could have been formed early or after the light mentioned in Gen 1:3 was formed. Most lights in nature seem to be emitted by stars. For instance, in the Solar System, most of the light that defines the day and night comes from the Sun. Likewise, the formation of light on the first day of creation implies that, by then, the precursor of the Sun was already radiating light, although that precursor could not have gotten its complete form yet.

Because they were still being modified, the precursors of the bodies other than stars (e.g., precursors of planets, asteroids, and satellites) could have also been emitting some light on Day 1. In other words, the light that was formed on Day 1 could have been visible not only from precursors of stars like the Sun, but also from the precursors of bodies that would not become a star. For, in the early moment of their formation, the precursors of many bodies could have been emitting light, or luminous, glowing, shining particles. The time it could have taken for photons (light particles) to form and shine bright enough to be seen may have been why the formation of light was not mentioned in the first verse of the creation narrative.

Those who think that the light mentioned in Genesis 1:3 was the glory of God made a mistake, for God cannot create His glory during the creation of the universe. He always has His glory. Likewise, because they do not understand that the precursor of the Sun was giving light before becoming the Sun on the 4th day, some people think that the glory of God was what gave light to the universe, or at least to the Earth, during the first 3 days of creation. Some people made this mistake because they think the Sun was spontaneously created on the 4th day, rather than formed from a precursor that underwent changes during the previous 3 days of the creation narrative. The precursor of the Sun could have been able to produce as much energy and light as the Sun does today, but was not molded or shaped into the spherical form of the Sun until the 4th day came.

Like I explained toward the end of this book, it is true that, in the heaven that will come down, the glory of God will be the light, and the Sun or Moon won't be needed indeed, but that glory of God was not the light that shone during the first 3 days of creation. The glory of God was neither the source of light on which the plants formed on the 3rd day, nor the source from which the Moon and Sun were formed on the 4th day. In other words, the light on which the plants formed on Earth on the 3rd day depended on was the light emanating from the precursor of the Sun.

Unlike the Earth, which has a crust hiding its luminous or volcanic interior, the Sun has no crust, and its light today could be similar to that of its precursors, which illuminated the first 3 days of creation. To sum up, the light of the precursor of the Sun was what the plants used on the 3rd day before the formation of the Sun was finalized on the 4th day.

Based on the evidence of today's day-and-night cycle, it is possible that about 12 hours could have passed from the time God began creating and forming things in the universe until He formed the light. Because God is light, He could not have been present in the darkness of the night. In other terms, God has never brought all His own light on Earth during creation, else no darkness should have been seen in the night. For His presence alone would have dissipated the darkness. Likewise, if

CHAPTER 17: UNLOCK THE CREATION OF DARKNESS AND THE FORMATION OF LIGHT

God had to deploy all of His force and energy to visit the Earth, or the Solar System, or even the whole universe, He could have easily destroyed it, just as a concrete block cannot fall on an egg without breaking it. In other words, it is also to preserve His creatures that God chose to stay on His throne in Heaven and never leave it. It makes sense that, to come on Earth as a man, Jesus, the Son of God had to "leave" His glory, some of His power, and other attributes in heaven. Then, after His mission on Earth was finished, He regained it again.

Electromagnetic fields are also expressions of particles similar to light. Visible light is just a manifestation of radiation in the visible spectrum of human beings. Some radiation that human beings cannot see is visible to certain wild animals. Likewise, plants are not sensitive to all forms of light, for they do not intercept all the light shining on them. The energy in photons comes from their precursors and the way they were packaged. The high speed of photons can partially explain the fact that they were not tightened together as other particles. In other words, composite particles could travel slower than light due to their clustering and the way they were formed. Because light is indispensable for life, God could have empowered light particles to move very fast, so they could also escape selfish human manipulations that might have affected them if they were meant to move slowly.

The bright white light is believed to consist of particles having various wavelengths. It is possible that the characteristics of the photons constituting the various forms of radiation could be different, meaning made of different particles having unique mass, speed, and wavelength. However, the scientific community seems to lump all photons into the same category and differentiate them solely by wavelength, without considering that their histories of formation could have been unique.

Despite the studies on light, many unknowns remain regarding its nature, whether viewed from a corpuscular or wave-like perspective. The errors made on the nature of light have been translated into scientific domains, including the explanation of the origin of celestial bodies. Because physical light is a metaphor for a key component of God, the human inability to fully understand it does not surprise me. The formation of light and of water on Day 1 suggested to me that many subatomic particles and composite particles could have been formed on the first day as well. It is possible that photons may not be elementary particles as existing theories allege, but composite particles, meaning they are made of other particles that, as of 2020, scientists are still struggling to identify.

One of the strategies that God designed to allow darkness to exist in the universe is to make light to have a near-linear trajectory or a curved trajectory with a very high (or infinite) radius rather than allowing it to move around all of the bodies it interacts with. In other words, if light was not designed to travel in a straight line or a near-straight line, it could not allow darkness to exist on some bodies like Earth in the night. For instance, while the face of the Earth that is in front of the Sun is hot and is in the light, the other side of the Earth is in darkness and is colder. This is because the light the Sun sends to the Earth travels almost in a straight line and cannot orbit the Earth to illuminate all its sides, including the side that is supposed to be in the dark.

The rotation speed and size of the Earth have been set to help regulate Earth's temperature and many other factors. Likewise, the darkness and cold in space also serve as a "brake" for the movement of light. Otherwise, light could be able to travel indefinitely. But due to the impact of cold and darkness, light particles can lose some of their energy, hence being capable of slowing down at some point during their journey in space. Imagine that the light emitted by all the galaxies in the universe is reaching us on Earth without being attenuated, or imagine that the light or radiation emitted by the Sun is reaching us without being cooled down! Life would be difficult or impossible. The fact that light can be slowed down or lose some of its energy is one of the reasons I believe scientists have made a huge mistake by assuming the speed of light is constant. The same assumption caused them to lose the measurements of the size and distance separating us from remote galaxies. In short, the characteristics of darkness and cold, and the trajectory of light, played a major role in the formation and maintenance of the universe. Praise be to God for His great design!

Genesis 1:3 says that *"And God said, let there be light."* The Hebrew word rendered as "light" is "**Yehi or**," "**vaihi or**," which signifies light as well as fire. That same word is also used for the Sun (e.g., Job 31:26, Isaiah 31:9, and Ezekiel 5:2), for lightning (Job 37:3), for the heat deriving from "esh", and for fire (Isaiah 44:16). This interpretation of the Hebrew word illustrates the amount of information encrypted in the Genesis 1 story. Some theologians have also said that Genesis 1:4 *(God divided the light from the darkness)* can be an "anticipation of the rotation of the Earth", which consequently divided the day (light) from the night (darkness).

17.2.2. The light that gives life

Jesus is the light of the world, hence just before His death, light went out, and darkness covered the land of Israel (and probably beyond) for 3 hours as a symbolic sign to the people: *"Now from the sixth hour* [noon] *there was darkness over all the land unto the ninth hour* [3 PM]. *About the ninth hour, Jesus cried with a loud voice, saying, Eli, Eli, lema sabachthani? That is to say, My God, my God, why have you forsaken me?"* (Matthew 27:45-46). This event implies that people cannot properly see certain things if they don't see them through Jesus. Hence, as He was agonizing, the Land of Israel symbolically went through a tumultuous darkness. I would not be surprised if the whole Earth were darkened at that time. In other words, without Jesus, the world walks in darkness. Jesus once said that anyone who follows Him will not walk in darkness: *"Then spoke Jesus again unto them, saying, I am the light of the world: he that follows me shall not walk in darkness, but shall have the light of life"* (John 8:12).

Jesus is also referred to as the bright and morning star: *"Jesus have sent mine angel to testify unto you these things in the churches. I am the root and the offspring of David, and the bright and morning star*" (Revelation 22:16). Apostle John also confirmed that Jesus Christ is the Light of the world which shone in the darkness, but the darkness comprehended Him not (John 1:1-12). In contrast, the physical darkness in our universe (the darkness of the night) seems to know and comprehend the physical light (e.g., the light of the Sun or the stars). However, the darkness in the heart of man does not comprehend the light of God. The main problem is about how human beings use their free will to see or to blind themselves.

There are many types of lights. Isaac Newton, one of the greatest scientists who ever lived, shed some light on the composition of light. Today, we know that a white light consists of many types of light. There are other types of light that science has not discovered yet. However, there is a type of light that gives life! Like I said earlier, Jesus said that He is the light of the world, and anyone that follows Him will have the light that gives life (John 8:12). We can deduce that every life draws its existence from a special light from God, which according to John, is the word of God that has been with God since the beginning (John 1). In contrast to the light of God that gives life, other types of "light" can kill. For instance, certain types of light that have more energy than X-rays can kill. Thanks to God that, to some extent, the Earth's atmosphere is shielded against these types of lethal light. The closer one gets to the Sun (or to most natural sources of light), the more they can hurt themselves. So far, human beings have not been able to get very close to the Sun yet. Scientists were able to get farther from the Sun by visiting the outer planets of the Solar System, but they have been unable to get very close to the Sun. This is because the Sun can burn them. Many artificial satellites sent closer to the Sun were destroyed because of the intensity of the Sun's light. Similarly, God never reveals himself to anyone physically. The Bible says that except for Jesus, no man has ever seen God, and that no human being can see Him and still live. God is light, and if a mortal and sinful man sees God's full glory in our current age, that person may die or have some problems. To meet the angel of God (that some people think was God himself), Moses had to cover his face because it shone so bright after being in the presence of God. At one point, when Moses asked to see the glory of God, God had to hide Moses in the cleft of the rock as He passed by. The glory of God is so bright that in the world to come, it will be the light, as the Sun will be no more. In the last days, God will darken the light of all stars (Ezekiel 32:8). On the day of the Lord, which some compare to the return of Jesus and/or the day of judgment, the Bible declares that the "*Son on Man [Jesus] will be like lightning that flashes and lights up the sky from one horizon to the other*" (Luke 17:24). Anyway, if Moses had really seen God, Jesus would not have said that no man has ever seen God. Some angels carrying the presence of God can act on behalf of God, and people experiencing them can think that they have seen God. Nevertheless, some people have visited heaven in the spirit and had experiences with God.

The light of Jesus gives life. Likewise, the light created by God gives life to this world, but a life that "is not" everlasting as that which Jesus gives. Without light, most life on Earth will stop. Plants need light to grow and produce leaves, fruits, and other things that men and most wild animals depend on. Without the light of the Sun, life could have stopped on Earth a long time ago. For instance, human beings get Vitamin D from sunlight, and Vitamin D deficiency can cause various health problems. I also understand that there are certain forms of beings on Earth which do not depend on or need light to live. So that I do not overload some people with too much scientific information related to those kinds of beings, I decided not to elaborate much on those organisms that live without input from Solar light.

Anyway, without the life of Jesus, human beings have no eternal life, although they may currently exist as beings that depend on physical light, such as that of the Sun. Hence, those who do not believe in God are like spiritually dead people walking in a temporary physical flesh denuded of the true life, which is the eternal life. In contrast, those who believe in and live for Jesus will live forever even if they may die. Although their flesh may die one day, their spirit will be alive forever. Just as the precursor of the Sun existed for 3 days before its shaping into the Sun was finalized on the 4th day, so also Jesus existed before the creation of the world and was then incarnated into a human being in the 4th millennium after the creation of the world. In other words, considering that 1000 years for men is like 1 day for God, and that Jesus was born to the Virgin Mary about 4000 years after Adam and Eve, this means He was born on the 4th day, just as the Sun was born on the 4th day of creation. Like I said before, according to the Jews, the day starts with the night. In other words, the formation of the Sun was completed in the second half of the 4th day. Likewise, the birth of Jesus occurred in the second half of the 4th millennium, meaning just before the beginning of the 5th millennium since Adam and Eve. The timing of the birth of Jesus was not by chance but followed a highly encrypted and divine plan involving human beings, some of whom acted as if they were the sole actors in the design God had worked out before laying the foundation of the world!

Just as the precursor of the Sun had to go through many changes, or generations of molding, or shaping, until the formation of the Sun was completed on the 4th day of creation, so also many generations of human beings descending from Adam and Eve, belonging to the genealogy of Jesus, were born, and had to live before Jesus was born, and God was working His redemption plan through them is just as a precursor being worked out to produce a final body. In other words, the birth of Jesus in the flesh through the Virgin Mary, about 4,000 years after Adam was formed in the Garden of Eden (on the 6th day of creation), was like the completion of the formation of the perfect human being, which pleased God. Jesus existed before creation began, of course, but it took God a little less than 4,000 years to work miracles through many generations, leading to the birth of Jesus. I praise God for the mystery hidden behind many things we take for granted.

Today, people see the Sun every day but they don't know that it symbolizes the life and grace of the Lord and is a testimony that God is still in control! Some people think that the Sun is God, therefore, just like the ancient civilization of the

Cahokia Mounds (a group of Native Americans who lived near St. Louis, Missouri, along the Mississippi River), they worshipped the Sun. Sunday was mistakenly dedicated to worshipping the Sun, while God Himself was rejected. The day is coming when the Sun will pass away and become dark. For those who worship the Sun, who will they worship on that day? I believe they will be disappointed on that day.

When He was on Earth as a human being for about 33 years, Jesus did not shine with the fullness of His glory. It was not by chance that Jesus started his ministry after he was 30, implying that, in a perspective where 10 years is 1 day, he began on the 4th day.

Some people think that before coming to the Earth, Jesus chose to leave His glory in heaven, others think that He brought His glory on Earth, while still others think that He used a limited version of His glory on Earth. Considering what I know about the origin of the universe, I felt like, to fulfill His mission as the Lamb of God and the Savior of the world, Jesus could not have brought or used all His glory to the Earth about 2000 years ago. Otherwise, He must have changed a lot of things on Earth and in human beings beforehand, just as He will have to do when He returns for the second time, as illustrated in the Book of Revelation. For, like I said earlier, just as a bulldozer or a tank cannot move over an egg without breaking it, God cannot come on Earth with all His power without destroying it before the due time. Otherwise, we are not talking about the Almighty God, strong in battle, the I AM, whose angels' presence the Israelites could not even stand in such a way that they had to beg God to leave. It is not by chance that God will have to destroy this Earth before bringing down heaven (the tabernacle of God)!

The issue of the past, current, and future state of the Earth and its inhabitants and even of the whole universe is not about what God can or cannot do, but about what His plan and timing. God can do anything He wants anytime of course, but based on His revelations, He does not act like that, for He is an extremely orderly master planner! Hence, those who do not know God's rules act contrary to His desires, while those who know them watch and pray for the due seasons and appointed times, which will never fail! I provide more details about these mysteries toward the end of this book. But the day is coming when those who will live in heaven will have no need of the light of the Sun for the glory of Jesus will be their light (Revelations 21:25). There would be no night in heaven, for the city needs no Sun or Moon as God will be our light (Revelations 21:23). The city also has no temple, as God Almighty and the Lamb (Jesus) are the Temple (Revelations 21:22). But now on Earth, we (Christians) are the temple of the Holy Spirit and the light of the world. Unfortunately, many are not shining as they should, which is not surprising; f,r Jesus Himself did not shine to His fullness when He was on Earth. Worse, some believers darken their light with sins.

The intensity of the light of God must be infinitely brighter than that of the Sun. Hence, no man can see God and still live. In the Old Testament, God sometimes hid in a thick cloud before speaking to the Israelites (Exodus 20:21), probably because He did not want to kill them with His light. When God called Moses to meet Him on top of Mount Sinai, God's glory covered the mountain for 6 days, and the Israelites saw the glory as a raging fire (Exodus 24:12-17). Nevertheless, Moses did not fear that fire, but he entered it to meet God and he stayed there for 40 days and nights (Exodus 24:18). Surely, God covered His glory with a cloud to prevent the Israelites from seeing it.

Several centuries later, Jesus Christ was born. One day, He took 3 of His disciples (Peter, John, and James) to pray on a mountain. At some point, the appearance of His face changed, and His clothing became gleaming white (Luke 9:27-29). Suddenly, Moses and Elijah appeared to Him in a glorious splendor. Around that time, the 3 disciples who were with Jesus fell into a deep sleep. As they woke from sleep, they saw the glory of Jesus and the two men (Moses and Elijah) standing with Him. Not knowing what they were seeing and saying, those disciples proposed to Jesus to build 3 tabernacles (one for Jesus, one for Moses, and one for Elijah). As they were speaking, God overshadowed and covered them with a cloud; and they were very fearful as they entered in the cloud (Luke 9:30-34). Then, God the Father told them to listen to Jesus. The Bible emphasized that, after that experience, the disciples never told anyone about it (Luke 9:35-36). God might have forbidden them to do so because Jesus did not want to die before His appointed time.

When Saul, who later became the Apostle Paul, met Jesus for the first time, he was blinded by the glory of God. Similarly, no one can look at the Sun for long without hurting themselves. Just as how human beings need to use special sunglasses before looking at the Sun, God used a cloud to shield humans from going blind if they see Him. When Jesus was born, He did not come to the world with His glory. But at the Mount of Transfiguration, His glory appeared, and when His disciples began talking about it and seeing it, Jesus was covered by a cloud to prevent the disciples from knowing the details of what was happening. And those who encountered God, if not the angels representing Him (for no mere human being has ever met God) were changed. For instance, because Moses stayed in the presence of God's glory for 40 days, his face shone (Exodus 34:29-35). Just think if Moses' face shone so much from being in the presence of God, how much brighter is God's face? I can only imagine.

God has a glory, and He does not want to show it to everyone yet. In the Old Testament, Moses and Joshua were allowed to enter that glory, whereas the other Israelites saw only a cloud from afar. When Jesus returns to rule on Earth, He will come in His glory (Matthew 24:29-31; Matthew 25:31-34). On that day, every tribe on the Earth will see Him, but some people will mourn as the Son of God will be coming down in the clouds of heaven with power and great glory to rule and judge the whole Earth and its inhabitants.

Today, God Himself dwells in the believers, enabling them to do great things in the natural. In contrast those who reject God get up early in the morning and go to bed very late without finding an answer for their true problem; for no darkness is as dangerous as the unbelief of those who reject Jesus Christ, the light of the world. As believers increase in their knowledge of God, their light will increase, and, although they will be living in physical bodies, they will have a greater influence and attract more things toward them, including blessings, power, and wealth just as a bigger body has the ability to exercise power due to its gravity. In other words, the impact they can make and the number of good things that believers can download from heaven or attract in this world can depend on the amount of the word of God (light) they have and apply to every aspect of their lives! I will expound on this in the chapter on gravity.

Finally, some Biblical references suggest that light could have played a role in the movement of some bodies in the universe. For instance, God talked to Job about a light that scattered the east winds upon the Earth (Job 38:24). According to the King James Version, the dispersion of light scatters the wind. Other Bible translations seem to separate the movement of the light from that of the wind. Nevertheless, knowing that the light that reaches the surface of the Earth can be reflected, and knowing that air particles can be moved, it will not be surprising that the reflection or dispersion of light scatters the winds. Knowing that winds can move things, the impact of light on winds suggests that light can move things in space, including even big bodies like planets.

17.2.3. Spiritual darkness and light

From a spiritual perspective, what people believe in or see with their spiritual eyes is what gives them their spiritual light. Because the light of the body is the eye (Luke 11:34), when the eye is evil, the body is also full of darkness (Luke 11:34). Therefore, when people accept God in their heart, God shines in their heart to give them the light of the knowledge of God's glory, which will show them the way of life (2 Corinthians 4:6), so they can live.

Just as darkness dominates where there is no light, sin also dominates where there is no light of God. Hence, those who live in sin are in deep darkness, while they may appear to be prospering physically. That is why when some angels rebelled against God, He chained them in darkness until the judgment day: "*For if God spared not the angels that sinned, but cast them down to hell, and delivered them into chains of darkness, to be reserved unto judgment*" (2 Peter 2:4). Other fallen angels are not chained but are ruling from dark places in this world, fighting human beings: "*For we wrestle not against flesh and blood, but against principalities, against powers, against the rulers of the darkness of this world, against spiritual wickedness in high places*" (Ephesians 6:12). In contrast, God is perfect and no darkness can be found in Him, for He is light: "*This then is the message which we have heard of him, and declare unto you, that God is light, and in him is no darkness at all*" (1 John 1:5). Moreover, He dwells with light, hence He cannot live in people who do not believe in Him. Those who think they know Him but who are walking in darkness, are lying: "*If we say that we have fellowship with him, and walk in darkness, we lie, and do not know the truth*" (1 John 1:6).

The more I studied the origin of the universe, the more I realized that God has hidden some mysteries, some of which are in Biblical parables and others are in plain sight, but people do not pay enough attention to figure them out. Although those mysteries are in plain sight, their real significance can be unearthed only when assessed from a spiritual perspective, pointing to a spiritual knowledge that relates to everything. As a result, those who refuse to believe in God will never understand certain things; hence, it is sometimes useless to try to explain certain things to certain people and force them to accept certain things they do not want to accept or they are not meant to ever understand because of the settings of their mind! That is why, despite the number of proofs some people will be exposed to about reality, they will never open their hearts to accept certain facts; instead, they will believe and accept what they think fits the formulas, models, or prejudices in their minds.

Just as darkness will never overpower light, meaning wherever light appears, darkness must disappear, so also the evil world can never overpower God, but wherever God appears, Satan and his demons must yield. For light has power over darkness. The temporary ability of the dark world to afflict some people bearing the light of God has caused some uninformed people to ignore the power of God, who very soon will imprison darkness and everyone living in darkness forever in hell, the only place where darkness will exist forever. Just as light comes with different intensities or brightness, so also do believers in God do not have the same light and consequently, they cannot have the same influence, discernment, or perception of the world.

As these thoughts raced through my mind on August 18, 2016, I felt like I was being enlightened to make some connection between some mysterious things in nature. That day, I felt like the relationship between darkness and light hides a secret that testifies about creation and eternity. Yet most human beings have not been paying attention.

Just as light and darkness cannot coexist in the same environment, God cannot coexist in the same environment as Satan and his demons. Just as the duration of the dawn (the time before the morning) and of the dusk (time before the night) is very short, so also is this age where light and darkness co-exist in the world is very short compared to the time when the eternal light (eternity in heaven) will never coexist with the eternal darkness (eternity in hell). Although in the world today darkness seems to be colder than heat, and the night usually colder than the day, the Bible reveals that the darkness in hell is very hot. In other words, the heat associated with the darkness of hell is to punish those who will be sent there, while the cold of the darkness of night serves another purpose, which I will explain later. I also later showed how the cold of darkness, which existed during the universe's formation, could have helped shape and influence some of the laws that governed creation. Although the light of God has no beginning, it is important to know that the darkness of Satan and his demons and even of human beings has a beginning. In the chapter on the formation of angels, I will explain how their fall induced an altercation between the spiritual and physical worlds. I could have addressed those details here, but I felt I should first address the origin or formation of the angels before the manifestation and consequences of their fall.

On December 29, 2017, it appeared to me that darkness played a crucial role in the creation of the universe and continues to play a role in maintaining its current state. Indeed, if the cold in space was smaller or higher during the formation of the celestial bodies, their precursors could have been shaped into different kinds of bodies. During the night for instance, the darkness and the cold that take place contribute to cooling down the Earth. This means that if the night was colder or hotter, the day-night temperature range could have been incompatible with the ecological niche of some living things known today, including human beings. To put it in a different way, the characteristics of cold and darkness agree with the laws that sustain life on Earth. The characteristics and reactions of some chemical elements could have changed if the variations of the conditions during the day and night had not been made to match life's requirements. For instance, if the Earth were as hot as Venus and Mercury, or as cold as Pluto, life would not have been possible, and Earth's chemical elements could have been different. The abundance of chemical elements varies from one planet to another, and more details can be found in my book on the origin of chemical particles. And if Earth were made to rotate slower than it currently does, the length of day and night could have been different and the characteristics of human beings could have been different or affected if they were made to match the existing environmental conditions.

According to the Bible, God created the Earth so it could be inhabited. At the beginning of the creation, God created darkness and the cold it contains. Then, He may have used its impact on the heat to control the formation of certain bodies in the universe. As I explain later in this book, the cold in some environments during the formation of the universe was indispensable for cooling and solidifying some precursors, including magma at the surface of the precursor of the Earth, so that a crust could form. In other words, if the temperature outside of the precursor of the Earth were as hot as that on the inside, the surface of the Earth would have never been transformed into a crust. Just as a freezer is needed to solidify water into ice, so also the darkness and the cold were needed during the formation of the Earth, and they are still needed to cool down the Earth.

Therefore, to fully understand the formation and functioning of the universe, special attention must be given to darkness and the cold. Although some people simply think that darkness is the absence of light and that cold is the absence of heat, God created them before light for a reason.

17.2.4. Stars and the moon will not shine light forever

Many prophets, including Joel, Ezekiel, Amos, and Isaiah, said that the day is coming when the heavens will tremble as the Sun and the Moon will stop shining:

- *"The earth shall quake before them; the heavens shall tremble: the sun and the moon shall be dark, and the stars shall withdraw their shining"* (Joel 2:10).
- *"The sun shall be turned into darkness, and the moon into blood, before the great and terrible day of the Lord come"* (Joel 2:31).
- *"And when I shall put thee out, I will cover the heaven, and make the stars thereof dark; I will cover the sun with a cloud, and the moon shall not give her light"* (Ezekiel 32:7).
- *"And it shall come to pass in that day, saith the Lord God, that I will cause the sun to go down at noon, and I will darken the earth in the clear day"* (Amos 8:9).
- *"The sun shall be no more thy light by day; neither for brightness shall the moon give light unto thee: but the Lord shall be unto thee an everlasting light, and thy God thy glory"* (Isaiah 60:19).

Jesus told His disciples that the darkness of the Sun will occur after the great tribulation and that God can transform physical light into darkness:

- *"Immediately after the tribulation of those days shall the sun be darkened, and the moon shall not give her light, and the stars shall fall from heaven, and the powers of the heavens shall be shaken"* (Matthew 24:29).
- *"The Sun shall be turned into darkness, and the moon into blood, before the great and notable day of the Lord come"* (Acts 2:20).
- *"The Sun shall be turned into darkness, and the moon into blood, before the great and notable day of the Lord come"* (Acts 2:20).

The Apostles also testified about the coming of that day when the Sun will go black: *"The sun shall be turned into darkness, and the moon into blood, before the great and notable day of the Lord come"* (Acts 2:20). God who called light out of the darkness in the beginning is also able to transform the physical light into darkness, therefore pointing to the power of God to reverse even the most complicated reactions.

Another Book by Nathanael-Israel Israel:
HOW GOD CREATED BABY UNIVERSE

THE FIRST AND ONLY BOOK THAT ACCURATELY EXPLAINS EVERYTHING ABOUT THE FORMATION OF THE UNIVERSE AND LIFE IN A WONDERFUL LANGUAGE THAT ALL CHILDREN AGES 7-12 CAN EASILY, FULLY UNDERSTAND & ENJOY!

As the only universe-origin book that your whole family will like and enjoy together, "*How God Created Baby Universe*" will set children on the path of success by accurately showing them early in life the formation of the universe, and how to detect errors in theories or stories that would misguide them as they grow up. Therefore, you need to add this great, efficient, trustworthy, and cost-effective book to the strategic journey of children toward their best tomorrow.

With "*How God Created Baby Universe,*" you will:

- Have peace of mind that children will get accurate, fitting, and easy-to-understand universe-origin information that will produce real results in their lives.

- Become the leader that captures the heart of children craving for the original explanation of the formation of the universe so you can clear their way for freedom, power, technology, innovation, and breakthroughs of the future (learn more at Science180.com/children)

- Protect yourself and loved ones from wrong theories in the literature and the media by keeping children secure and empowered with the true knowledge of how the universe began

- Explain complicated secrets to children about how to locate mistakes in origin-related theories so you can save time, money, and other resources to improve their lives

- Ultimately boost children's confidence in detecting, confronting, and avoiding wrong theories by knowing the facts and real processes involved in the formation of the universe

- Help children to easily sort out their origin-related questions using strategies that get them to tap into deep secrets that even highly educated people ignore

- Clearly explain to children how to mathematically know without a doubt whether God created the universe as the Bible says or billions of years of evolutionary processes formed it

- Accurately explaining the complex formation of the universe and of life to children can be very hard in our modern world, but by getting *"How God Created Baby Universe"*, you will know the proven formula to help children to easily understand their huge universe-origin and life-origin questions with confidence, humor, and joy.

They will surely laugh aloud while reading this book and thank you for it! It is time to buy this pragmatic book to help the children in your life today.

A member of the American Association for the Advancement of Science, American Chemical Society, and the American Society for Microbiology, **Dr. Nathanael-Israel Israel is** a Beninese-American scientist and international consultant, who shows the world how to scientifically decode the formation of the universe, of life, and who is known as the creator of the Chemicals Turbulent Origin Formula™, the inventor of the Life Turbulent Origin Formula™, and the discoverer of the Universe Creation Formula™. Learn more at Israel120.com.

Another Book by Nathanael-Israel Israel:
HOW BABY UNIVERSE WAS BORN

If you don't believe in God or you hate God, or you don't think there is anything or anyone called "God," but you want your children to understand how the universe was formed from a scientifically-proven perspective that considers the facts, then this book is for your children. This book can be a very good gift to an unbeliever you want to bring to God.

Dr. Nathanael-Israel Israel is the founder of Science180, the American organization that helps people enter the realm of true knowledge about the universe's formation. In other words, he is known as the first human being to use modern science to provide people with a state-of-the-art decoding of the origin of the universe and of life.

CHAPTER 18

WHY YOU DON'T HAVE TO BELIEVE ANYONE TALKING ABOUT THE FORMATION OF GALAXIES AND CHEMICAL PARTICLES, INCLUDING WATER, BEFORE DISCOVERING A TOP SECRET HIDDEN HERE

Most bodies in the universe are composed of chemical particles, and before my discoveries, no comprehensive theory had explained their origin. Therefore, I cannot talk about the origin of the universe without mentioning chemistry. To understand the formation of chemical particles, I relied on scientific data on their characteristics, including abundance. But because I devoted a book to the formation of chemical particles, here, I will not detail how they were formed. For the sake of time and space, I will just review a few things about chemical elements: name, origin, abundance, and state at standard temperature and pressure.

18.1. Name of the chemical elements

Chemical elements were named after various things, such as a city or country, a famous scientist, a planet, an asteroid, an idol, a god, or a goddess. While some names are rooted in Greek, others are from Latin, German, Japanese, Spanish, Swedish, and Old English. To help readers understand the origins and meanings of the names of chemical elements, I have provided some examples below. In parentheses, I put the chemical element symbol and its number of protons, Z (a number used to estimate the mass or weight of atoms). The higher that number, the heavier the chemical element associated with it. The origin of the names of the chemical elements is based on the work of Yinon (2016).

Chemical elements named after a city or a country include:

- Americium (Am, Z=95): After America
- Berkelium (Bk, Z=97): After Berkeley, California
- Californium (Cf, Z=98): After California (State and University)
- Copper (Cu, Z=29): After the island of Cyprus
- Darmstadtium (Ds, Z=110): After Darmstadt, Germany, where the element was first synthesized
- Dubnium (Db, Z=105): After Dubna, Russia
- Erbium (Er, Z=68): After Ytterby, a town in Sweden
- Europium (Eu, Z=63): After Europe
- Francium (Fr, Z=87): After France
- Gallium (Ga, Z=31): After the Latin word Gallia, the old name of France
- Germanium (Ge, Z=32): After the Latin word Germania, meaning Germany
- Hafnium (Hf, Z=72): After the Latin word Hafnia (Copenhagen)
- Hassium (Hs, Z=108): After the Latin word Hassias, a German state
- Lutetium (Lu, Z=71): After Lutetia, an ancient name of Paris
- Magnesium (Mg, Z=12): After the city Magnesia in Europe
- Moscovium (Mc, Z=115): After Moscow Oblast, Russia, where the element was first synthesized
- Polonium (Po, Z=84): After Poland
- Rhenium (Re, Z=75): After Rhines, a province of Germany
- Scandium (Sc, Z=21): After Scandinavia
- Strontium (Sr, Z=38): After Strotian, a Scottish town
- Tennessine (Ts, Z=117): After Tennessee, a state in the USA, where Oak Ridge National Laboratory is located
- Terbium (Tb, Z=65): After Ytterby, a town in Sweden
- Thulium (Tm, Z=69): After Thule, the ancient name of Scandinavia
- Ytterbium (Yb, Z=70): After Ytterby, a town in Sweden
- Yttrium (Y, Z=39): After Ytterby, a town in Sweden

Chemical elements named after a famous scientist include:

- Bohrium (Bh, Z=107): After Niels Bohr, a Danish physicist
- Copernicium (Cn, Z=112): After Nicolaus Copernicus, a Polish astronomer
- Curium (Cm, Z=96): After Pierre and Marie Curie, the pioneers of radioactivity
- Einsteinium (Es, Z=99): After Albert Einstein, the author of the relativity theory
- Fermium (Fm, Z=100): After Enrico Fermi, a famous scientist after whom

the subatomic particles called fermions are named

- Lawrencium (Lr, Z=103): After Ernest Lawrence, a physicist
- Meitnerium (Mt, Z=109): After Lise Meitner, an Austrian physicist
- Mendelevium (Md, Z=101): After Dmitri Ivanovitch Mendeleyev, the chemist who pioneered the chemical elements table named after him.
- Nobelium (No, Z=102): After Alfred Nobel, the one who founded the Nobel Prize awards
- Oganesson (Og, Z=118): After Yuri Oganessian, a Russian physicist
- Roentgenium (Rg, Z=111): After Wilhelm Conrad Röntgen, a German physicist
- Rutherfordium (Rf, Z=104): After Lord Rutherford, a New Zealand chemist and physicist
- Seaborgium (Sg, Z=106): After Glenn T. Seaborg, the scientist who discovered many transuranium elements

Some chemical elements were named after a planet:
- Mercury (Hg, Z=80): After the planet Mercury
- Neptunium (Np, Z=93): After the planet Neptune
- Plutonium (Pu, Z=94): After the planet Pluto
- Uranium (U, Z=92): After the planet Uranus

A chemical element was even named after an asteroid:
- Cerium (Ce, Z=58): Ceres, a main belt asteroid

Just as for some celestial bodies, many chemical elements were named after an idol, god, or goddess:
- Niobium (Nb, Z=41): After Niobe, daughter of the mythical king Tantalus
- Palladium (Pd, Z=46): After Pallas, a Greek goddess of wisdom and after an asteroid
- Promethium (Pm, Z=61): After Prometheus, a god who is believed to have stolen fire from the sky and gave it to man
- Tantalum (Ta, Z=73): After Tantalus, a mythological Greek king
- Thorium (Th, Z=90): After Thor, a Scandinavian god
- Vanadium (V, Z=23): After Vanadis, a Scandinavian goddess

Elements that were named after a German word include:
- Bismuth (Bi, Z=83): After the German word wissmuth (meaning white mass)
- Cobalt (Co, Z=27): After the German word kobalt or kobold (meaning evil spirit)

- Nickel (Ni, Z=28): After the German word kupfernickel (meaning false copper)
- Zinc (Zn, Z=30): After the German word zin (meaning tin)

Examples of chemical elements named after Greek words are:
- Actinium (Ac, Z=89): After the Greek word aktinos (ray)
- Antimony (Sb, Z=51): After the Greek words anti (opposed) and monos (solitude), hence "not alone"
- Argon (Ar, Z=18): After the Greek word argon (inactive)
- Arsenic (As, Z=33): After the Greek word arsenikos and the Latin word arsenicum
- Astatine (At, Z=85): After the Greek word astatos (unstable)
- Barium (Ba, Z=56): After the Greek word barys (heavy)
- Bromine (Br, Z=35): After the Greek word brômos (stench)
- Cadmium (Cd, Z=48): After the Greek word kadmeia (ancient name for calamine) and from the Latin word cadmia
- Chlorine (Cl, Z=17): After the Greek word khlôros (green)
- Chromium (Cr, Z=24): After the Greek word chrôma (color)
- Dysprosium (Dy, Z=66): After the Greek word dysprositos (hard to get at)
- Helium (He, Z=2): After the Greek word hêlios (sun)
- Hydrogen (H, Z=1): After the Greek words hudôr (water) and gennan (generate)
- Iodine (I, Z=53): After the Greek word iôdes (violet)
- Krypton (Kr, Z=36): After the Greek word kryptos (hidden)
- Lanthanum (La, Z=57): After the Greek word lanthaneis (to lie hidden)
- Lead (Pb, Z=82): After the Greek word protos (first)
- Lithium (Li, Z=3): After the Greek word lithos (stone)
- Molybdenum (Mo, Z=42): After the Greek word molubdos (lead)
- Neodymium (Nd, Z=60): After the Greek words neos (new) and didymos (twin)
- Neon (Ne, Z=10): After the Greek word neos (new)
- Osmium (Os, Z=76): After the Greek word osmë (odor)
- Oxygen (O, Z=8): After the Greek words oxus (acid) and gennan (generate)
- Phosphorus (P, Z=15): After the Greek words phôs (light) and phoros (bearer)
- Praseodymium (Pr, Z=59): After the Greek words prasios (green) and didymos (twin)
- Protactinium (Pa, Z=91): After the Greek word protos (first)
- Rhodium (Rh, Z=45): After the Greek word rhodon (rose)

- Selenium (Se, Z=34): After the Greek word Selênê (Moon)
- Technetium (Tc, Z=43): After the Greek word technêtos (artificial)
- Tellurium (Te, Z=52): After the Greek word tellus (Earth)
- Thallium (Tl, Z=81): After the Greek word thallos (young shoot)
- Titanium (Ti, Z=22): After the Greek word titanos (Titans)
- Xenon (Xe, Z=54): After the Greek word xenon (stranger)

A chemical element named after a Japanese city is:
- Nihonium (Nh, Z=113): After the Japanese city Nihon, a place in Japan where the element was first synthesized

More than 12 chemical elements were named after a Latin word:
- Aluminum (Al, Z=13): After the Latin word alumen
- Calcium (Ca, Z=20): After the Latin word calcis (lime)
- Carbon (C, Z=6): After the Latin word carbo (coal)
- Cesium (Cs, Z=55): After the Latin word caesius (sky blue)
- Fluorine (F, Z=9): After the Latin word fluo (flow)
- Holmium (Ho, Z=67): After the Latin word Holmia (Stockholm)
- Iridium (Ir, Z=77): After the Latin word iridis (rainbow)
- Manganese (Mn, Z=25): After the Latin word mangnes (magnet)
- Radium (Ra, Z=88): After the Latin word radius (ray)
- Rubidium (Rb, Z=37): After the Latin word rubidus (red)
- Ruthenium (Ru, Z=44): After the Latin word Ruthenia (Russia)
- Silicon (Si, Z=14): After the Latin word silex (flint)
- Sulfur (S, Z=16): After the Latin word sulfur (brimstone)

A chemical element named after a Spanish word is:
- Platinum (Pt, Z=78): After the Spanish word platina (little silver)

A chemical element named after a Swedish word is:
- Tungsten (W, Z=74): After the Swedish words tung sten (heavy stone)

The chemical elements named after Old English words are:
- Gold (Au, Z=79): After the Old English word geolo (yellow)
- Silver (Ag, Z=47): After the Old English word seolfor (silver)

As a Christian, I was a little annoyed, surprised, and uncomfortable that no chemical element was named after God the creator, Jesus, Yahweh, Adonai, El Shaddai, Elohim, Jehovah, which are some of the names of the true and only God who created everything. I already explained how even the scientific evidence agrees with the Biblical story of creation. I pray that human beings can better honor God when they deal with the creatures, and that some key constants, equations, and facts be named after the name of God, the Creator, not after idols. It is time that scientists start honoring God, the creator, instead of the creations of man.

18.2. State of the chemical elements at the standard temperature and pressure

The state of a matter is the form in which it can exist. Matter is usually classified according to 4 states and Wikipedia (2020) characterized them as:

- Solids have a fixed volume and shape, with particles close together and fixed in place.
- Liquids have a fixed volume but a variable shape that adapts to fit their container, and their particles are close together, but move freely.
- Gases have a volume and shape that adapt both to fit their container, while their particles are neither close together nor fixed in place.
- Plasmas have variable volumes and shapes and contain particles which can move around freely.

Usually, by changing the pressure and temperature, the states of chemical elements can be changed, suggesting that one of the players in the formation of chemical elements could have been a process which created different small pockets of space occupied by different precursors of subatomic particles, atoms, and precursors of other types of matter to be spiraled, rolled, shaped, squeezed, and compressed differently.

The environmental conditions that shaped, squeezed, and compressed the types of matter could not have allowed gases to be compressed like liquids or liquids to be compressed like solids. The differences among these states of matter could have been affected by the positions and turbulence of their precursors. If the conditions were met, free electrons in plasma, which, that some people believe were ripped away from their nuclei, are electrons that would not have initially orbited any nuclei in the beginning. They are like secondary bodies that were formed without being able to properly orbit a primary nucleus, resulting in a normal atom. As I explained in the section on the formation of the stars, the turbulence intensity was lower than that required to complete the formation of some rocky planets. The small intensity of the turbulence in the stars could not have allowed the aggregation of all their constitutive particles into a cluster of primary subatomic particles (e.g., nucleons) orbited by secondary subatomic particles (e.g. electrons). That is why stars like the Sun are believed to be made of plasma. Regions where the turbulence could have been higher in the precursors of those stars could have allowed the formation of heavier elements, but due to the nature of the turbulence of the precursors of stars, their main constituents are particles dispersed in plasma. The intensity of the turbulence that led to the formation of the Earth and its particles did not fit conditions that favored the presence of pure plasma; hence, a plasma state is said to not freely exist under normal conditions on Earth, but could be commonly generated by other processes like lightning, electrical sparks, etc. Under Earth-like environmental conditions, particles in plasma could be transformed into normal particles and chemical elements similar to those found on Earth. It should not be surprising that the particles that the Sun "projected" into space could be transformed into other particles when they reached environments having a turbulence level different from that on the Sun. Other types of subatomic particles and atoms not found on Earth could be present in the plasma of stars, for the level of turbulence could have allowed the aggregation of subatomic particles into other kinds of atoms not found in an environment dominated by higher turbulence. Similarly, in environments where turbulence could have been stronger and particles more tightly packed, denser subatomic elements and atoms could have been found, even denser than those currently known on Earth. In other words, it is a mistake for scientists to believe that chemical elements are the same from one stellar system to another and from one galaxy to another.

The atoms and subatomic particles of the precursors of each state of matter were gathered differently and brought close together at different speeds, depending on the conditions of their formation. The atoms and molecules in solids were brought closer together than those in a fluid. Consequently, atoms of solids are held together more tightly than those in liquids. The intermolecular cohesive forces in fluids are not sufficient to hold their constitutive atoms together. Hence, fluids generally flow under the influence of some stress and instability. Molecules or atoms in gases are much farther apart than those in liquids. This makes gases more compressible, and they can also expand indefinitely when the external pressure is removed. Although liquids are considered to be relatively incompressible, during the formation of the universe, their precursors were greatly compressed until a certain "limit" or equilibrium was reached. However, if the proper amount of pressure is applied to any liquid, I believe that a liquid can be compressed and the nature of its atoms or molecules can change.

18.3. Mystery and formation of water

Water is first mentioned in the Bible in the second verse of the Biblical account of creation: *"And the earth was without form, and void; and darkness was upon the face of the deep. And the Spirit of God moved upon the face of the waters"* (Genesis 1:2). A few verses later, a firmament, or an expanse, or a sky was formed in the midst of the waters as the waters were divided or separated from the waters (Genesis 1:6). Three verses later, it is mentioned that the waters under the heaven were gathered together unto one place and the dry land appeared (Genesis 1:9).

Water was used to form the creatures that have life, and fowl that fly above the earth in the open firmament of heaven (Genesis 1:20). However, the etymology of the Hebrew word "shamayim" which is the heavens referred to in Genesis 1:1 suggests to me that water could have been formed as early as the first verse in the Bible. For "shamayim" implies a body containing water or a water-like compound. In other words, the formation of water could have been quick. Because the Bible did not detail how water was formed does not mean that water has always existed.

Jesus (the fountain of living water) would later reveal that, "Verily, verily, [...] except a man be born of water and of the Spirit, he cannot enter into the kingdom of God" (John 3:5), therefore signifying the crucial importance of water not only for physical things, but also for spiritual ones. Jesus also said " ... whosoever drinketh of the water that I shall give him shall never thirst; but the water that I shall give him shall be in him a well of water springing up into everlasting life" (John 4:14). In John 7:38, he continued by saying: "He that believeth in me, as the scripture hath said, out of his belly shall flow rivers of living water". This is one of the verses some people use to assume that the water mentioned in Genesis 1 came or flowed from the belly or the throne of God. But nowhere in the Bible is it mentioned that water flowed from God's belly.

After carefully reviewing the scriptures and scientific data, I do not think water existed before the creative events mentioned in Genesis 1, contrary to what some people think. For, being a physical thing, water had to be formed by God. However, in the spiritual realm, I also understand that Jesus, the living water, existed before creation, but He is not the water created in Genesis 1. In other words, the spiritual water that Jesus symbolizes was not the physical water mentioned in Genesis 1.

On November 18, 2013, I got the preliminary ideas about how water could have been formed. Indeed, water on Earth is found in 3 forms: gas, liquid, and solid. In the Solar System, it sounds like the Earth is the main planet where water is abundantly present in all 3 forms. At one point during Earth's formation, water covered the surface and had to be removed by God so dry land could appear. To my knowledge, no other celestial body in the Solar System is as abundant in all 3 forms of water as Earth. Water was found on other celestial bodies including the Sun. While on celestial bodies closer to the Sun, water could have evaporated due to solar radiation, on those farther from the Sun, water could have been frozen in gaseous or liquid form. Meteorites are also said to be rich in water. Consequently, some people wrongly believe that water was brought to Earth by comets. What a mistake about the origin of water! If that was the case how come they aren't still watering Earth?

Because as it is mostly known, life on Earth requires more chemical compounds than just water, the presence of water on other celestial bodies is not enough to claim that they contain living organisms. If living things were found on other celestial bodies, they would be different from those found on Earth. For things present in each celestial body are meant to match the characteristics of those places. That is also why those who plan to move to Mars or the Moon and live there are just joking, for they do not understand that such a move threatens their very existence!

Water is a fluid indispensable for the existence of human beings and most other living things on Earth. Water played a crucial role in some Israelite spiritual rituals, including sprinkling it and washing people and objects for purification. In the book of Leviticus, water is referenced a lot. During the Israelites' journey back to Israel after their stay in Egypt, God provided water for them from a Rock (Numbers 20:11), which would later be revealed by the Apostle Paul as Jesus Christ, the Rock of Ages. The scriptures mention how Jesus baptized those who believe in Him not with water like John the Baptist, but with the Holy Ghost (Ruach ha-Kodesh) described as fire (Acts 11:16). In heaven, a pure river of the water of life will proceed out of the throne of God and of the Lamb (Revelations 22:1) and whosoever will, will take that water of life freely (Revelations 22:16-18) as part of the biblical rituals that will last forever and ever. A long story short, water is crucial for life, and no human being can live without it, and any environment that lacks water is unsuitable for most forms of life!

I cannot finish this chapter on water without saying a few words about the birth of rain on Earth. Indeed, from the creation of the Earth until the great flood during Noah's time (Genesis 7), no rain had ever fallen onto the Earth. For instance, during the days of Adam and Eve, a mist from the Earth watered the Garden of Eden: *"But there went up a mist from the earth and watered the whole face of the ground"* (Genesis 2:6). It was after the great flood (in the days of Noah) that water started raining.

Considering the description of the layers of water in Genesis 1, I felt that the water that rained during the great flood could have been part of the water that moved above the Earth on the second day of creation. And just as He did during the flood of Noah, God could have ordered a wind to blow and cause the water covering the Earth to go down, inside the ground or into the sea: *"And God remembered Noah, and every living thing, and all the cattle that were with him in the ark: and God made a wind to pass over the earth, and the waters assuaged"* (Genesis 8:1). In other words, during the formation of the Earth, God could have used a wind to remove the water covering the Earth, then opened the ground so it could go underground into the seas, or under the crust, or in rivers and lakes found on the continents. Others could have been pushed and stacked in the icebergs in the north and south poles.

On a different note, past and future events show that water can be transformed into blood, meaning that God is able to transform any chemical compound into others. Indeed, in the events preceding the departure of the Israelites from Egypt to Canaan, the first plague inflicted on the Egyptians was God using Moses to turn the Nile River into blood (Exodus 7:19-21). In the future, streams of water will also be transformed into blood (Revelations 16:4). This miracle of changing water to blood indirectly points to the role that water played during the formation of the universe and also to the power of God who can create and change anything into anything.

Although I do not want to discuss them here, according to other Jewish books,

God formed water after spreading light over the darkness present at the beginning of creation. This suggests that some complex particles could have been formed after simpler ones. More details about the formation of water can be found in *"Origin of the Spiritual World."* I encourage you to check it out to unearth more mysteries and enlighten your understanding of creation.

Finally, the formation of water by the first day suggested that many other chemical compounds could have been formed by then. In other words, because water is a chemical compound, the data of its formation suggests that other chemical compounds and elements could have been formed by then, while others would have still been going through processes finalizing their formation. Knowing that the formation of the Earth was finalized on Day 3 and that of the Sun on Day 4, I deduced that most chemical elements could have been formed by Day 4. Because plants were formed on Day 3 and some animals on Day 4, I deduced that most chemical elements found in plants and animals could have been formed by Day 4. Because of the factors involved in properly explaining the formation of chemical elements, I will revisit the formation of water in the next chapter and discuss it further in my book on the origin of chemical particles.

18.4. Abundance and formation of water

For instance, the abundance of a chemical element measures its occurrence relative to all other elements in a given environment. In my book on the origin of chemical particles, I devoted many pages to the abundance of chemical elements in the universe, our galaxy, the Sun, human beings, oceans and seawater, Earth's crust, planets in the Solar System, and the atmosphere and crust of the Moon. But here, I will just summarize the abundance of water and the atoms (oxygen and hydrogen) which constitute it.

The Bible did not say much about how water was formed. Some people even consider water as having always existed, and others think that it emanated from the throne of God. However, because it is a chemical compound (a substance formed from two or more chemical elements), water was formed by God, not created. In other words, water was made from material God created out of nothing.

Chemically speaking, a water molecule (annotated H_2O) is made of 2 atoms of hydrogen (H) and one atom of oxygen (O). In the next subtitle, I explain how water was formed. Moreover, in my book on the origin of chemical particles, I showed that the precursor of water was split-gathered into the precursor of hydrogen and the precursor of oxygen; and as they were molded by turbulent processes, these precursors formed water. The formation of water on the first day of creation does not mean it was formed from nothing. The steadfastness of the formation of water should not cause us to forget the discipline or order that God followed during creation. He could not have made water and then broken it down into its constituent atoms.

CHAPTER 18: DECODE THE FORMATION OF WATER, GALAXIES, AND OTHER CHEMICAL PARTICLES

Water is vital for life and without it, life cannot exist on Earth. Before I start writing about water, let me recall here that, in addition to the ordinary water which is made of oxygen and the ordinary hydrogen, the atmosphere of some planets (e.g. Mars) contains another type of water called "heavy water" (annotated ^2H_2O or D_2O) because it is made of oxygen (O) and the hydrogen isotope deuterium (^2H or D), instead of the ordinary hydrogen-1 isotope (^1H or also called protium).

NASA (2018) proved that water is also present in the atmosphere of the 4 giant planets as water ice. Water is found in 3 forms in the atmosphere of the planets: liquid water, water ice, and heavy water (i.e., Hydrogen-Deuterium-Oxygen). NASA reported that liquid water was found in the atmosphere of all 5 innermost planets in the Solar System: Mercury, Venus, Earth, Mars, and Jupiter. The atmosphere of no planet in the Solar System is as rich in water as the Earth's atmosphere, where water accounts for 1% of its volume. Water is very abundant on Earth, in the oceans, rivers, lakes, living organisms, etc. Variations in environmental conditions across locations and seasons can obviously affect atmospheric water content. This also suggests that evaporation alone cannot explain the high concentrations of certain constituents in the atmospheres of some planets.

Seeing the water content of the atmosphere of the planets, it sounds like the atmospheres of Mercury and Venus were not suitable to have a very high water content probably because of the high temperature and other environmental conditions. Additionally, Earth's conditions are optimized to maintain high water content. However, as the precursors of the planets got farther from the Sun than the Earth, the conditions got harder for water to stay in the atmosphere. That is why the water content in the atmosphere decreased from Earth to Jupiter passing by Mars. Beyond Jupiter, the conditions could not have been suitable for the formation of liquid water in the atmosphere. Hence, it is frozen and is in a form of aerosol.

It is important to mention that, as of 2020, scientists have found at least a trace of water in almost all of the celestial bodies they studied. Some satellites in the Solar System are even thought to have liquid water. Even in the Sun, scientists have discovered water. This evidence points to the existence of water at an early stage of the formation of the Solar System. If this is true for the Solar System, it may also be true for other stellar systems in the universe, therefore confirming the Biblical account of creation in Genesis 1 according to which water was present in the beginning. However, what people have missed is that God formed the waters mentioned in Genesis 1 using the initial matter He created. Furthermore, it is important to recognize that the waters in Genesis 1 could have contained other chemical particles.

Due to the stage of human knowledge in the days of Moses, Moses could not have used any better words than "waters" to describe the nature of the fluids of the precursors of celestial bodies including the Earth. In the days that Moses wrote the Book of Genesis (about 1450-1410 BC according to the King James Version of the Bible), little was known about chemistry. Even about 1000 years after Moses, when Greek philosophers such as Empedocles (c. 490–430 BC), Plato (c. 428-347 BC), and Aristotle (384–322 BC) tried to classify matter, they mostly used 4 terms (earth, water, air, and fire). Additionally, Aristotle added a fifth term: aether. It took more than 3,000 years after Moses for most chemical elements to be discovered.

Using about 56 chemical elements, the Russian chemist and physicist Dmitri Ivanovich Mendeleev (1834-1907) formulated the periodic table of chemical elements around 1863. As of 2020, more than 118 chemical elements have been discovered and characterized; some are artificial. Among these 118 chemical elements, only nine were known before the Common Era (meaning before the birth of Jesus):

- Sulfur (S, Z=16)
- Iron (Fe, Z=26)
- Copper (Cu, Z=29)
- Silver (Ag, Z=47)
- Tin (Sn, Z=50)
- Antimony (Sb, Z=51)
- Gold (Au, Z=79)
- Mercury (Hg, Z=80)
- Lead (Pb, Z=82)

Again, the letter Z that I mentioned in the symbols of the chemical elements are connected with their mass or weight—the higher that number, the heavier the element. Besides those 9 chemical elements known for more than 2000 years, the discovery year of the other elements I studied ranges from 1400 to 1996. But some synthetic chemical elements whose properties are still under scrutiny were discovered more recently, after 1996. Besides the above 9 elements, only 4 others were discovered by the end of the 17th century:

- Bismuth (Bi, Z=83): 1400
- Zinc (Zn, Z=30): 1500
- Phosphorus (P, Z=15): 1669
- Carbon (C, Z=6): 1694

Then, in the 18th century, 19 chemical elements were discovered. In the 19th century alone, meaning from 1801 to 1899, fifty (50) additional chemical elements were discovered. In the 20th century, 30 elements were discovered. If Moses had used some modern scientific terminology in the creative narrative, his contemporaries would not have had any clue about what he meant. If, despite what we know today, most top scientists are failing to decode the meaning of the Biblical account of creation reported by Moses, then similar tasks could have been scientifically more difficult in those days, when science was not yet established. Yet, people have believed in Moses and many still believe in the Genesis 1 story even today. For God's plan is for people to believe in Him instead of to scientifically demonstrate Him before "believing."

In short, Moses did an outstanding job in recounting the revelation he received from God concerning the creation and formation of the universe. No mere human being could be so accurate without God revealing things to him. The processes I studied regarding the formation of the bodies' precursors (small and large) in the universe also confirmed that the story told in Genesis 1 is very accurate. Although Moses did not discuss many chemical elements, scientific evidence suggests that as water was forming, other chemical elements were also being formed. Unfortunately, because they do not understand the formation of the universe, particularly the Biblical account of creation in the first chapter of the Bible's Book of Genesis, some people (including educated believers) think that water was the raw material that God used to form every other chemical particle in the universe. But water was formed using the initial matter that God created in the universe. Because water is composed of hydrogen and oxygen, among the known chemical elements, I decided to review just these 2 to give a glimpse of how water was formed, hoping that those who want the full demonstration can refer to my book on the formation of chemical particles.

18.5. Abundance and formation of hydrogen (H, Z=1) and hydrogen-based compounds

The lightest nonmetal in the atmosphere of the planets is hydrogen (Fig. 16). The two isotopes of hydrogen found in the atmosphere are the following:

- Protium: the ordinary hydrogen atom that has one proton and one electron
- Deuterium: the hydrogen isotope that has one proton, one neutron, and one electron

Deuterium is the least abundant isotope of hydrogen in the atmosphere of the planets, where it is present in 2 forms:

- Hydrogen Deuteride (HD) is found in the atmosphere of the 4 giant planets
- Hydrogen-Deuterium-Oxygen (HDO), also called heavy water, is only found on Mars

In contrast, the ordinary hydrogen is found in combination with other atoms in the form of 12 molecules including ethane (C_2H_6), hydrocarbons (C_2H_x), methane (CH_4), methane ice (an aerosol), molecular hydrogen (H_2), ordinary water (H_2O), water ice (aerosol), hydrogen-deuterium-oxygen also called heavy water (HDO), Hydrogen Deuteride (HD), ammonia (NH_3), ammonia ice (NH_3) which is an aerosol, and ammonia hydrosulfide (NH_4)HS), an aerosol.

Molecular hydrogen (H_2) is the most dominant form of hydrogen in the atmosphere of celestial bodies. For instance, molecular hydrogen accounts for 80% to 96.3% of the atmosphere of the 4 giant planets (Fig. 16). In nature, hydrogen associates with other elements. For instance, besides being found in the form of hydrogen gas (H_2), hydrogen in the atmosphere of the planets is present in other forms in various compounds such as: ammonia (NH_3), ammonia hydrosulfide, ammonia ice (aerosols), hydrocarbons, ethane, hydrogen Deuteride, hydrogen deuterium oxygen, methane, methane ice, water, and water ice (aerosols). The presence and abundance of each of these compounds depend on the planets. All of these compounds are not present in the atmosphere of all planets.

Fig. 16: Abundance (%) of **Hydrogen** (H, Z=1)

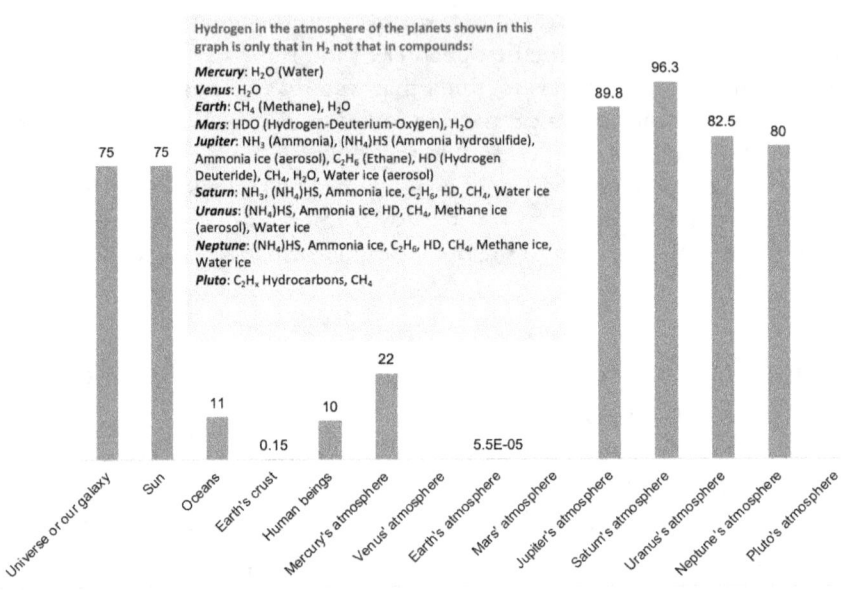

In general, hydrogen is more abundant in the atmosphere of the giant planets than in the Sun, the Earth's atmosphere, or Earth's crust. On Earth, it is more abundant in the ocean than in the crust. In general, its abundance is minimal in the atmosphere of most terrestrial planets except that of Mercury. Hence, hydrogen is more abundant in the environment of bigger bodies (e.g., giant planets and the Sun), but it can also be found in environments where heavy metals are found (e.g. Mercury's atmosphere). I felt like its abundance is defined by the level of turbulence and the other chemicals in its environment. When the energy and movement in some precursors of atoms were not strong enough to form heavy and dense atoms, hydrogen was abundantly made. After heavier elements were made, hydrogen could also have been produced from the leftovers of the precursors of the atoms that could not be molded into heavier or denser atoms.

It is not by chance that Mercury's atmosphere is the only atmosphere in the Solar System where metals have been significantly found. Of course, metals are abundant in the crust of most planets, but Mercury's atmosphere is the only atmosphere that contains metals. The abundance of the precursors of hydrogen and oxygen in the environment defined the formation of water. Some atmospheres that are abundant in hydrogen today lack a significant amount of water because oxygen was limiting. Similarly, although the atmosphere of the 4 giant planets is rich in hydrogen, they are poor in water and some are even deficient in water because they are not rich in oxygen. It is not that hydrogen and oxygen were formed separately and then joined to form water; rather, their precursors were formed in compartments that allowed them to combine to form water.

Another chemical compound usually formed with hydrogen is methane (CH_4). Methane exists in the atmosphere of all of the planets from Earth to Pluto except Mars. The lack of methane in the atmosphere of Mars may be due to the fact that the precursors of the chemical particles were preferably oxygen and carbon, which formed the carbon dioxide (CO_2) occupying more than 95% of Mars' atmosphere, while other carbon atoms associated with other oxygen atoms formed carbon monoxide (CO) in Mars' atmosphere. In other words, the conditions of the precursor of Mars and the Martian atmosphere could have favored the dominant formation of carbon and oxygen, which are bound together to form CO_2, rather than carbon binding with hydrogen to form methane. The few hydrogen atoms that were formed in the atmosphere of Mars bound with some oxygen atoms to form water (H_2O) which is just 0.021% of the Martian atmosphere while the remaining hydrogen atoms on Mars were associated with deuterium (a hydrogen isotope) and oxygen to form the heavy water (Hydrogen-Deuterium-Oxygen) which as of 2020, is significantly found in the atmosphere of Mars and in the atmosphere of the other planets in the Solar System.

18.6. Abundance and formation of oxygen (O, Z=8) and oxygen-based compounds

Based on the data I studied, the highest abundance of oxygen in the Solar System was found on Earth, where it dominates the oceans by 86%, the Earth's crust by 46%, and the Earth's atmosphere by 20.95% (Fig. 17). Human beings are 61% made up of oxygen. The atmosphere of Mercury is the only other place where a higher abundance of oxygen (42%) has been found in the Solar System.

Considering the crucial role that oxygen plays in the physiology of human beings and other living organisms, it is important to underline that the Earth is a privileged place to host human beings and we are blessed to live on Earth. Some scientists are trying to find living organisms on other planets and even on other stellar systems, but their efforts will conclude that, life on Earth is not by chance. Therefore, we need to be grateful to God, the Creator, for placing us in an environment that matches our needs. Otherwise, if human beings were made in any of the other environments I studied, life could already have been extinguished, for those environments do not contain sufficient amounts of the chemical elements required to power and sustain the biological reactions and chains that sustain life.

According to the environment it is found in, oxygen associates with other chemical elements to form various compounds. For instance, in the Earth's atmosphere, oxygen atoms are associated with carbon atoms to form CO_2, or associated with hydrogen to form water (H_2O). Oxygen atoms can also bind to one another to form the oxygen gas (O_2), that living organisms breathe or to form the ozone (O_3), which is said to shield the Earth from some toxic radiations coming from the Sun or other bodies in space. In the atmosphere of the planets, oxygen is present in different compounds:

- Carbon Dioxide (CO_2),
- Carbon Monoxide (CO),
- Hydrogen Deuteride (HD),
- Hydrogen Deuterium Oxygen (HDO),
- Nitrogen Oxide (NO),
- Sulfur Dioxide (SO_2),
- Water (H2O), and
- Water ice (Aerosols).

In the atmosphere of the giant planets, the dominance of hydrogen and the lack of oxygen atoms could have limited the formation of water. Hence the small amount of oxygen in the atmosphere of the 4 giant planets is sequestered like aerosols in those giant bodies where the atmospheric pressure is also very high.

The fact that oxygen which is the second most abundant element in the Earth's atmosphere is the #1 most abundant element in the Earth's crust suggests that there was higher turbulence or harsher conditions in the precursor of the Earth, which yielded other denser atoms that could have favored its abundance in the crust. And because many precursors of atoms were differentiated into oxygen in the precursor of the Earth's crust, and others into heavier atoms, a small portion was left to be converted into smaller atoms. Hence the Earth's crust is rich in heavy atoms, but lacks some lighter ones. Consequently, nitrogen, which was abundantly formed in the Earth's atmosphere could not be abundantly formed in the Earth's crust, for the conditions in the precursor of the Earth's crust were suitable for the formation and abundance of atoms heavier and denser than nitrogen.

Oxygen is very indispensable for the life of most animals, particularly those containing hemoglobin. On Earth, oxygen is the second most abundant gas, accounting for 20.95% of the atmosphere. However, the planet whose atmosphere is the richest in oxygen is Mercury (42%) (Fig. 17). Venus is denuded of a significant amount of oxygen (O_2), whereas on that planet, oxygen atoms were combined with carbon atoms to form a huge amount of CO_2. In contrast, the oxygen content of Mars's atmosphere is 0.13%. That's a big difference between Mars and the Earth.

The low oxygen content of Mars implies that, although some people dream of colonizing Mars with human beings, the oxygen supply may be a problem coupled with the high toxicity of carbon monoxide as explained earlier. Do these people think that they will not need oxygen to live on Mars or do they think they can produce enough to solve their needs in the long run? Although Mercury may contain more oxygen than Earth, its temperature is sometimes very high (more than

400 K, which is about 136° C) and sometimes very cold (below freezing) when it is not exposed to the Sun, making it impossible for humans to live there. This means that the Earth occupies a privileged position in the Solar System, offering maximum protection from heat while providing enough oxygen and CO_2 for animals and plants.

I therefore thank and praise God for the oxygen supply on Earth so that we, the human beings, can breathe freely. I also thank Him for not filling the Earth with toxic gases like carbon monoxide. I must also give Him thanks for making the Earth's temperature manageable for us human beings, other nonliving things, and living organisms to exist, survive, and still enjoy the variations of the seasons.

In addition to the ordinary water which is made of oxygen and the ordinary hydrogen, the atmosphere of some planets (e.g., Mars) contains another type of water called "heavy water." Beyond Jupiter, the conditions could not have been suitable for the formation of liquid water in the atmosphere. Hence, it is frozen in the form of aerosol. This also explains why comets are rich in ice, for they are located far away from the Sun and the fluids in their precursors were frozen.

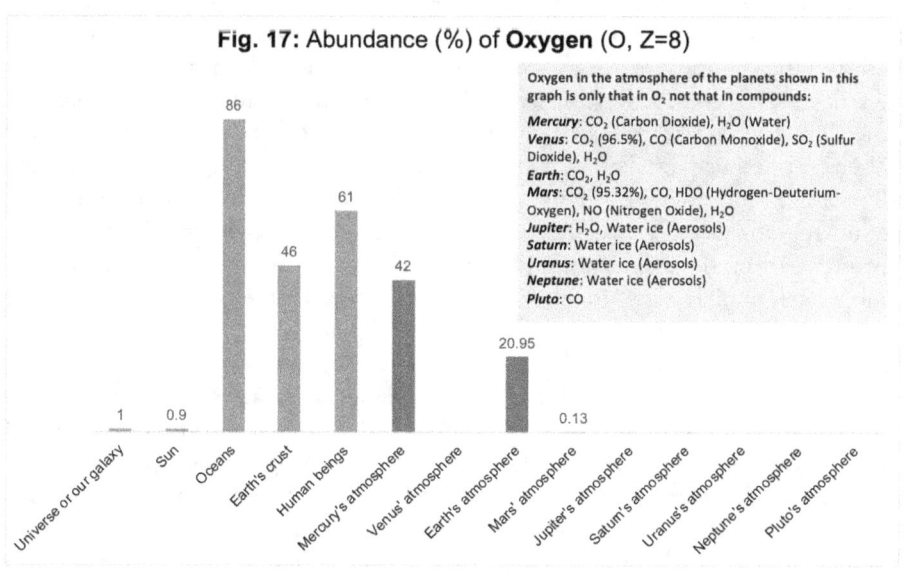

Fig. 17: Abundance (%) of **Oxygen** (O, Z=8)

Oxygen in the atmosphere of the planets shown in this graph is only that in O_2 not that in compounds:

Mercury: CO_2 (Carbon Dioxide), H_2O (Water)
Venus: CO_2 (96.5%), CO (Carbon Monoxide), SO_2 (Sulfur Dioxide), H_2O
Earth: CO_2, H_2O
Mars: CO_2 (95.32%), CO, HDO (Hydrogen-Deuterium-Oxygen), NO (Nitrogen Oxide), H_2O
Jupiter: H_2O, Water ice (Aerosols)
Saturn: Water ice (Aerosols)
Uranus: Water ice (Aerosols)
Neptune: Water ice (Aerosols)
Pluto: CO

18.7. Timing of the differentiation of atoms

Atoms and subatomic particles could not have been differentiated before the precursors of the celestial bodies were split. Likewise, the chemical elements found in the atmosphere could not have been just a product of evaporation from the crust. For instance, some chemical elements not found in the atmosphere are very abundant in planetary crusts. For instance, silicon (Si, Z=14), which is not found in the atmosphere of the planets in the Solar System, is the second most abundant element in the Earth's crust and probably in the crust of other planets. The truth I unearthed concerning what could have happened in the atmospheres of the planets also gave me a glimpse of what could have happened during the formation of the planets' crusts. To put it another way, considering the trends I found in the formation of atoms in the atmosphere, I felt that the environmental conditions in the precursors of the planets' crusts could also have affected their chemical composition.

It appeared to me that all atoms and subatomic particles might not have been completely differentiated before the precursors of the celestial bodies and their clusters were split. For instance, if the atoms in the planetary systems of the Solar System had been established before the precursor of the Solar System was split into the precursors of the planets, the atmosphere of the giant planets could not have been dominated by hydrogen and helium, which are in very small amounts in the atmosphere of the terrestrial planets. Similarly, the atmospheres of the terrestrial planets could not have been dominated by heavier elements, whereas those of the 4 giant planets are dominated by lighter elements. The lack of oxygen in the atmosphere of the 4 giant planets, the Sun, or in other parts of the universe may be limited by the conditions that the precursors of atoms in those bodies were subjected to. The connection between the factors that shaped the celestial bodies and their atmosphere better explains the small density of the biggest celestial bodies and their dominance by lighter elements, while the smallest celestial bodies are usually denser and filled by denser elements. The processes that dominated the gathering together of the precursors of the bodies have also significantly affected the differentiation, development, and characteristics of their chemical elements. The same logic can explain why the chemical composition of the Sun is closer to that of the giant planets than it is to that of the terrestrial planets.

Likewise, the environment where the crust was formed was different than that of the atmosphere. Consequently, for the same planets, the abundance and nature of the chemicals formed in the crust are different than those in the atmosphere. For the Earth for instance, the crust is dominated by oxygen and silicon while the atmosphere is dominated by nitrogen and oxygen. Yet, nitrogen is just a trace in the Earth's crust, while silicon is barely present in (if not absent from) the Earth's atmosphere. In general, nonmetals (noble gases, halogens, and the conventional nonmetals) are more abundant in the atmosphere and waters than in the crust of the Earth. In contrast, metals and metalloids are more abundant in the crust than in the waters and the atmospheres. The difference of the environmental conditions where turbulence shaped the precursors of the particles in the crust, ocean, and atmosphere explains the difference, presence, and dominance of certain groups or types of chemical elements according to their locations.

I found that forming complex compounds may have been harder than forming simple ones. More energy is required to convert the precursors of heavier chemical elements into heavy atoms than the precursors of lighter atoms into light elements.

CHAPTER 18: DECODE THE FORMATION OF WATER, GALAXIES, AND OTHER CHEMICAL PARTICLES

That is why many precursors were converted into lighter or simpler chemicals than into heavier or more complex chemicals. In the same manner, the energy in the precursors of bodies (small and large) after they split from their mother could have also affected how their shaping and characteristics were "finalized." The scope of the changes the daughter bodies underwent after their split can explain the similarities and differences among their peers born from the same mother precursor.

For instance, the heaviest chemical elements are more abundant in the atmosphere of the innermost planets in the Solar System because such atmospheres favored the formation of dense particles. The conditions that compressed the precursor of the planets to increase their density also affected the nature of the atoms present in their atmosphere. It is also possible that some compounds in the atmosphere of the planets might have volatilized from the crust after their formation. However, most chemical elements in the atmosphere are more likely to be formed there directly rather than having migrated from elsewhere (e.g., the crust) into the atmosphere. In other words, some elements found in the atmosphere were formed there and did not just migrate there after being formed somewhere else. Some atoms in the atmosphere could have been formed under conditions different from those in the crust. For instance, while nitrogen is 78% of the Earth's atmosphere, in the Earth's crust it is just 0.02%, implying that conditions in the precursor of the Earth's crust could not have favored the abundant formation of nitrogen as conditions in the precursor of the Earth's atmosphere did. If all of the nitrogen atoms in the atmosphere were formed in the crust before migrating to the atmosphere, the abundance of nitrogen in the crust could have been higher and more traces of nitrogen could have been found in the crust. Yet, nitrogen is less abundant in the Earth's crust than many lighter gases.

Another way to explain this is that, as the celestial bodies were forming, in addition to some chemicals which were formed in the atmosphere, others might have escaped the crust or evaporated to position themselves above the crust or the more condensed part of the outer portion of these bodies. Those that evaporated could have later cooled down, while others might have fallen to the ground due to gravity and other processes. Others could never have fallen but could have stayed in the atmosphere. Some chemicals could have been ejected into the atmosphere as the celestial bodies were being wrapped into their shape. Other chemicals in the atmosphere could have been formed in situ after some of the precursors of their peers could have condensed and incorporated into the main bodies in their vicinity, and ended up forming structures like rings.

Using the formula of the chemical compounds present in the atmosphere of the planets, I was able to understand how the environmental conditions during the formation of different matters affected the split of their precursors into various entities and how the resulting products could have combined or bonded to others to yield different forms of matter (e.g., particles, atoms, molecules, and compounds) according to the scale of their precursors. For instance, in the Earth's atmosphere, some oxygen atoms combined with atoms of other chemical elements to form

Science180: Source of Unconventional Wisdom and Knowledge on the Creation of the Universe, Life, and Chemicals different compounds. Indeed, some oxygen atoms are associated with carbon to form carbon dioxide (CO_2). Other oxygen atoms are associated with hydrogen atoms to form ordinary water (H_2O). At a lower altitude, oxygen atoms associated with themselves to form oxygen gas (O_2), while at a higher altitude, they associated with one another to form the ozone (O_3). On some planets like Mars, the precursor of hydrogen not only yielded the ordinary hydrogen (also called protium, which has one proton and one electron), but also another isotope of hydrogen such as "deuterium" (which has one proton, one neutron, and one electron). As atoms were mixing in some environments (e.g., Martian atmosphere), some atoms of protium, deuterium, and oxygen associated to yield the heavy water known as hydrogen-deuterium-oxygen. When some lava was ejected into the atmosphere of the precursor to the planets, it could also have carried chemicals that remained there.

The time chemical particles can also explain the similarities and differences of their abundance according to the celestial bodies. To illustrate, I will briefly talk about the similarities between the chemical composition of the Moon and Earth. Indeed, according to scientific evidence, not only do the Moon and the Earth appear to have similar bulk compositions, but they also share virtually the same isotopic fingerprint. This similarity can be explained by the fact that the Moon and the Earth descended from a common precursor: the precursor of the Earth-Moon system. Like I explained in previous chapters, after the precursor of the Earth-Moon system was split into the precursor of the Earth and the precursor of the Moon, the precursors of their atoms and subatomic particles may not have gone through significant enough changes that could erase the footprint of their common origin and characteristics already imparted into them. Probably, many chemical elements in the Earth and Moon would have been completed or could have reached a significant stage of differentiation before the precursors of these celestial bodies were split, or the differentiation of the precursors of their particles could have reached a level of similarity that could not be reversed by the processes which finalized the formation of each of these celestial bodies. Consequently, although some differences exist between the chemical composition of the Earth and the Moon, the core of their characteristics and isotope composition is said to be almost similar. In other words, the similarities of the chemical composition and isotopic fingerprint of the Moon and the Earth can be explained by the fact that they both descended from a common precursor (the precursor Earth-Moon system) before each of them was changed after their separation. In *Turbulent Origin of the Universe*, I also explained why Jupiter's composition seems globally similar to that of the Sun. More details about the origin of chemical particles can be found in the book I wrote on that subject.

18.8. Underlying causes of the diversity of celestial bodies

The diversity of the bodies in the universe can be traced back to the conditions that

CHAPTER 18: DECODE THE FORMATION OF WATER, GALAXIES, AND OTHER CHEMICAL PARTICLES

prevailed during their formation. Based on facts that I already demonstrated in this book, I felt like I should elaborate a little bit on how the processes that molded the precursors of the celestial bodies affected the outcome of their daughter bodies. Hence, some celestial bodies are stars, others are gas planets, others are ice planets, others are rocky planets, etc. Why is all that?

The size of the precursors of celestial bodies affected the fate of their internal particles and overall state. For instance, because most of the precursors of their particles were not converted into very dense particles and because of their huge size, the precursors of stars were not able to have their outer surface solidified as a crust, meaning that stars lack an outer solid shell. Stars are believed to be mostly made of hydrogen and helium. At the planetary level, the precursors of some planets were able to form dense bodies and a crust that covers their outer part. That is the case for the terrestrial planets such as Mercury, Venus, Earth, Mars, and Pluto. For instance, I established earlier that, during the formation of the Earth, rocks were formed from the precursor of the Earth, and then piled to form the Earth's crust, while the inner portion of the Earth kept the fluid-like state. Therefore, the Earth sometimes erupts through volcanoes when an opening is made in the crust or when lava forces its way through. The superposition of the fluid layers in the precursors of celestial bodies explains why rocks are deposited on top of one another. Some fluids in the precursors of terrestrial planets acted as the "solvent" that glued some rocks and their constituents. While the interior of most planets would contain fluids, in the case of many small asteroids and satellites, no fluid could be found, only stacks of rocks. For the smaller the precursor of the bodies, the higher the likelihood that the bodies could have lost their internal heat. In contrast, the precursors of the bodies of some giant planets, such as Jupiter and Saturn, were not able to convert the precursors of their atoms into dense particles, which could have formed a dense crust. Instead, they formed a giant gaseous planet. In contrast, the precursors of Uranus and Neptune ended up forming icy planets. Between these planets, asteroids of various chemical compositions were formed. Depending on their position, some asteroids are believed to have a chemical composition intermediate between the metal-rich terrestrial planets and the volatile-rich outer bodies. Moreover, because of the circumstances of their formation, the main-belt asteroids were not gathered into a single asteroid or planet by piling their rocks, but instead the rocks were scattered and became individual asteroids organized into a belt. In other words, in the case of the formation of the terrestrial planets, the rocks were not scattered into individual celestial bodies, but they were heaped together and bound by "forces," including gravity. Beyond the asteroid main belt, because conditions did not favor the formation of a solid crust, the giant planets are dominated by gas. It is possible that some of the giant planets have a crust surrounded by a thick atmosphere. However, as of 2025, to my knowledge, no mission has ever collected samples from the surfaces of the giant planets (Jupiter and Saturn are gas giants, and Uranus and Neptune are ice giants). Sending a man-made mission to those would be very expensive and very risky. Unlike the precursor of the main belt asteroids, where at least rocks were formed without being able to gather together, for the precursors of the giant planets, rocks could not have even been able to form. Instead, the particles that were supposed to be packed together into rocks were loosened and spread out over a larger volume, thereby giving the giant planets a larger size.

Because it was very big, even bigger than the precursor of any other body in the Solar System, the precursor of the Sun was not much compressed or packed by the forces that shaped it to form a solid mass. Instead, most of its constituents or particles were less dense and were not as compressed as the precursors of some other bodies in the Solar System. The precursor of the Sun was like a big light or a very hot body filled with energy. The nature of the Solar System's precursor explains why the centers of Earth and most planets could contain lava. As the precursors of the celestial bodies were moving farther from the precursor of the Sun, the fluids in their precursors cooled down because of the decrease in temperature. This explains why the exteriors of the terrestrial planets are hard, whereas their interiors are likely similar to lava or magma that sometimes erupts from volcanoes.

Because of its massive size and intense light, the Sun emits a lot of heat into its surroundings and has never experienced a night, yet remains very hot and is not affected by the cold of the space around it. In other words, the precursors of the other bodies in the Solar System were also initially very hot like the precursors of the Sun, but because of the location they were brought to and the turbulence they went through, their outer surface was not able to preserve its state of matter and temperature, but was cooled down by the cold and darkness of the space they were projected into. Consequently, the outermost surface of some bodies was solidified, forming a crust atop a likely fluid interior. In contrast, the outer surface of some bodies was not solidified, but formed gases, while others formed icy surfaces. For instance, the environment and the nature of the turbulence in the precursors of Jupiter and Saturn explain their gaseous state, as their particles could have been highly agitated and more dispersed rather than packed into a solid planet. Furthermore, the precursors of the outer bodies, which are farther from the Sun, had to endure colder environments that removed a significant amount of their energy. Consequently, the surface of some of them was frozen. The precursors of the innermost bodies may have lost less energy because the Sun's precursor could have raised the temperature of their environment. In other words, the environment where the outer bodies in the Solar System were formed could have sucked a lot of energy from their precursors. Even today, some bodies could still be losing more energy due to their location. In other words, in addition to the turbulence intensity, which defined the nature of the particles in the precursors of the bodies, the temperature of their surroundings also played a role.

However, I think that in some locations of the galaxies, it is possible to find stars that are smaller than some planets in the Solar System. As long as the environment is very hot and clustered with many stars, its size would not matter. If the environment is very hot and cannot allow the cold to cool down the surface of the precursor of a star, it will form a star regardless of its mass. In the galactic core, where many stars are clustered, smaller stars can be found, meaning that size alone cannot justify the formation of a star.

Photons of light are matter that has been contracted, compacted, and packed together. Therefore, photons, which are the particles of light, freely move at a high speed. Because the particles in the Sun are not tightly tied together, some of them, including photons, are easily released. However, planets and satellites have their constituent particles tightly bound and do not easily escape or move out. Because the precursors of the particles in the bodies that are not stars were combined with others to form particles more complex than photons, they do not exist in the form of light. The energy in particles that are not photons somehow bears witness to their original formation based on something similar to what was used to form light.

Just as the chemical composition is not the same at every location on Earth, so also the chemical composition of the Sun and of other celestial bodies cannot be expected to be the same. For instance, although the Sun is dominated by hydrogen and helium, it also contains heavy atoms, although in smaller amounts. Heavy elements are also present in the Sun because when the precursors of matter in the Sun were being molded into atoms, the environmental conditions in some spots or pockets allowed their split-gathering into heavy elements instead of the smaller and lighter hydrogen or helium. The light and other forms of radiation that the Sun emits at a specific zone can depend on the type of chemical elements present in that zone. Some sunspots could be dominated by heavy chemical elements, which, therefore, decrease the brightness of the Sun at their location. The darker sunspots may contain denser chemical elements than the Sun's brighter spots. The variation and the random appearance of the localization of a sunspot may be due to the movement of the plasma in the Sun. As the Sun is moving, the matter that constitutes it keeps moving as well.

In summary, stars are like naked celestial bodies lacking the crusty cover that some celestial bodies (e.g., satellites, terrestrial planets, and asteroids) have. On the other hand, celestial bodies that have a crust are like clothed bodies whose nakedness (fluids in their interior) is covered by a crust. Lava inside the Earth proves that its precursor was very hot and was cooled down to form a crust outside. The lava or magma inside some planets and satellites indicates that their constituents did not undergo the changes that could have solidified them. But as seen with volcanoes, the lava or magma can solidify. The Sun has no crust because its intense heat burns away anything that could form one.

Rings found around planets are bodies I could discuss at this point, but because I already devoted many pages to them in *"Turbulent Origin of the Universe,"* I will not dwell on them here, but I will just summarize some processes related to their formation. Indeed, rings are discs of solid materials (e.g., dust, moonlets, and other small objects) orbiting an astronomical object. Although many planets, asteroids, and even satellites in the Solar System have rings, Saturn's rings are the best known and most frequently cited in the literature. In my book *"Turbulent Origin of the Universe,"* I devoted many pages to rings. When the precursors of a planetary system split-gathered, it yielded (1) the precursor of the primary planet, and (2) the precursor of its satellites and rings. As the fluid layers of the precursor of the satellites and rings began to experience turbulence, some particles coalesced into rocks of various sizes, while others coalesced into much smaller particles. The particles that were gathered together into larger bodies became the satellites, while those that were not incorporated into the precursor of the satellites became those that altogether formed the rings. In other words, during the formation of the rings and satellites, there was a point where their precursors were particles that were split-gathered into the precursors of the rings and the precursors of the satellites as their turbulence was underway. As turbulence was developing and fluids were moving, mixing, rotating, spiraling, and packing, some fluids were packed into bigger clusters, which yielded satellites, while other fluids did not amass into bigger bodies, but into smaller particles spread over the space occupied by their mother precursors. The daughter bodies, which were not assembled into satellites, became the particles constituting the rings. In contrast, the gathering together or condensation of the particles' precursors into denser bodies led to the formation of the satellites. Hence, some satellites are embedded inside rings.

18.9. Formation of galaxies

Stars are the most visible things in the sky to the naked eye. Some are believed to be hundreds of times wider than the Sun. While some stars in the universe are isolated, others are organized into clusters (e.g., open and globular clusters), and still others are arranged into galaxies and galaxy clusters. Although in *"Turbulent Origin of the Universe"* I devoted dozens of pages to characterizing stars and galaxies, here I will focus on a few aspects of their formation and organization.

Most galaxies are organized into structures called groups, clusters, superclusters, and clusters of superclusters. The structure of the galaxies is defined by 3 components: spheroidal component, disk component, and spiral arms. I showed that the precursors of some galaxies were rolled over, tilted, or inclined during their formation, and consequently, their rotational axis changed. Just as some celestial bodies in the Solar System do not have a system of primary and secondary bodies, some galaxies also may lack some components. Just as some bodies orbit others in the Solar System, within some galaxies, stars are said to orbit the bulge, also called the galactic center.

Unlike systems of bodies in the Solar System, where primary bodies are clearly separated from their secondary bodies, on the galactic level, the gathering together of the precursor of some galaxies could not have been strong enough to release or separate the precursor of the galactic arms farther away from the spheroidal component. The size and the degree to which the precursors of the spiral galaxies pushed away, ejected, or released their secondary stars away from the primary stars (which constitute the spheroidal component) could have determined the properties of the arms such as the size of the pitch angles, the angle of the extension around the center, the closeness or openness, and the number of complete rotations around the center of the arms.

According to their visual morphology, galaxies are classified into 3 main types: elliptical, spiral, and irregular. The precursors of normal spiral galaxies could have had less "difficulty" ejecting and split-gathering their secondary stars than barred spirals. For the stars that are in the arm are secondary, while those in the nucleus can be perceived as the primary star. The precursors of barred spiral galaxies could not have been able to eject their secondary stars directly away from the primary star, but they could have flown for a while before spiraling. At one point, the precursor of the barred region in barred spirals could have been almost in the middle of the arms, organized as if one were leading and the other trailing. As the precursor of the barred spirals wound up, because it could not wind up all of its stars to form an elliptical galaxy, the leading or trailing arms were tilted or wound up according to the direction or sense of the precursor's movement. The central part of the stars could not have been compressed a lot or wound up much because of their thickness, resistance to winding up, and the lack of energy that could have been required to compress them or spiral them.

As the precursors of galaxies were split-gathering, stars were being formed until it came to a point when a split could no longer occur or continue to occur due to the size of the precursors and the environmental conditions. At that moment, the formation of new stars could have stopped. The precursor of each star was wound up to yield an individual star. According to their environments, some stars were tightly wound up while others were loose. The degree of tightness and turbulence in the precursors of stars could have defined their internal composition and, consequently, the brightness of their daughter stars.

In *"Turbulent Origin of the Universe,"* I showed that the precursors of galaxies were astronomical fluid layers, which, due to their turbulence, swirled or were curled or wrapped into 3D bodies. The precursors of some galaxies were tightly collected into spherical bodies, forming, for instance, elliptical galaxies. In other cases, the fluid layers of the precursors of some galaxies were not compacted into elliptical galaxies, but into spiral galaxies. The central part, or core, or spheroidal component of spiral galaxies could have been where the fluid layers of their precursors were mostly trapped during the processes that brought them together. As the gigantic fluid layers of the precursors of the galaxies were being gathered together, stellar systems were formed inside them. Fig. 18 is a sketch of some configurational or conformational changes that the precursors of some galaxies may have undergone.

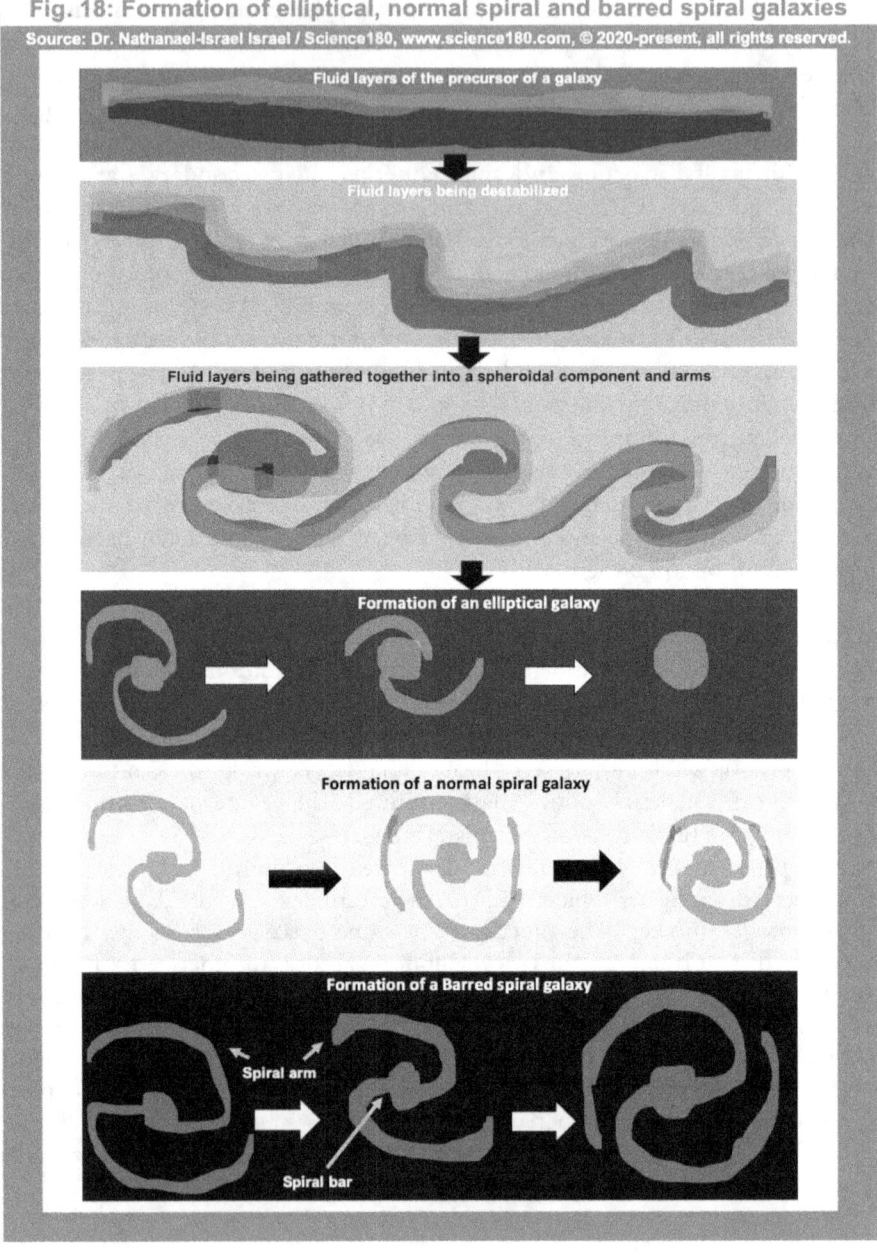

Fig. 18: Formation of elliptical, normal spiral and barred spiral galaxies

Nathanael-Israel Israel: The Scientific Prophet of Prophets

CHAPTER 19

THINKING THAT ALBERT EINSTEIN AND ISAAC NEWTON REALLY EXPLAINED GRAVITY? WHAT IF IT CAN BE SCIENTIFICALLY PROVEN THAT THEY WERE ALL WRONG … BECAUSE THEY MISSED THIS?

19.1. Meaning of gravity

Gravity is a parameter that expresses the ability of celestial bodies to attract matter that is close to their vicinity. Gravity is believed to be what causes things to fall on Earth when they are thrown into the air. Throughout the ages, people have formulated theories to explain gravity, but I have been satisfied with none of them. Therefore, I cannot write a book on the formation of the universe without addressing gravity, given my insight into this complex phenomenon. In this book, I will not delve into the gravity details that I handled in other books, but I will briefly explain what I think its cause is.

As of 2025, besides my discoveries, which explained gravity differently, two main theories of gravity exist:

- The gravitational theory of Isaac Newton was based on a hypothetical "action at a distance", and
- The relativity theory of Albert Einstein is based on a hypothetical "curvature of space"

Indeed, Isaac Newton postulated that the force (between 2 bodies) associated with gravity is proportional to the product of the mass of the bodies and inversely proportional to the square of the distance between them. In contrast, in his general theory of relativity, Albert Einstein viewed gravity as a consequence of the curvature of spacetime.

RECONCILING SCIENCE AND CREATION ACCURATELY

Despite the scientific efforts made to explain gravity, many questions still remain. For instance, while addressing the mechanics of gravity, Richard Feynman (a laureate of the Nobel Prize in Physics) said:

> *"What about the machinery of gravity? All we have done is to describe how the Earth moves around the Sun, but we have not said what makes it go. Newton made no hypotheses about this; he was satisfied to find what it did without getting into its machinery. No one has since given any machinery."* (Feynman, 2006).

Based on the extensive evidence I discussed in my book titled *"Turbulent Origin of the Universe"*, I showed that:

- gravity is not responsible for the orbit of the celestial bodies, and it does not control the trajectory of the bodies
- gravity is not the process that caused the secondary bodies to orbit their primary bodies, and it is not a force of attraction acting between all matter
- gravity is not responsible for the mass, the structure, and the development of the universe
- gravity is not responsible for the configuration and constitution of atoms and subatomic particles

Considering the data I gathered on the universe, it seemed to me that gravity relates to the impact of vortical structures (on all scales) on the fluid flow of the precursors of celestial bodies during their formation. Gravity is the consequence of the vortices of the precursors of the celestial bodies during the turbulence that formed the universe. Then, after the formation of the universe, the impact left a gravitational field around each celestial body. The way the fluid layers and the vortical structures in them were moved, sheared, revolved, rotated, tilted, and elongated had aggregated some precursors into bigger bodies and others into smaller ones. Before some bodies reached their size, fluids in their surroundings accumulated as their precursors moved until they reached a position where their volume could no longer be sheared any farther by the fluid flow. As the bodies' precursors were being rotated, gravity progressively formed according to the rotation and size of these bodies.

Considering the processes that explain the formation of the universe, I came to realize that gravity is a consequence of the split-gathering of the precursors of the celestial bodies into their daughter bodies. Without a mechanism, process, or force to combine the precursors into unified bodies, most daughter bodies could not have formed after their split from their mothers. The process of gathering matter into the precursors of bodies was not initially controlled by gravity. It was later that gravity itself was born. Another way of explaining this is that, before what is known today as gravity was established, forces and interactions were already acting on the precursors of bodies, helping gather their matter into unified bodies and systems of bodies. Gravity was like the mature or last state of expression of some of the forces that acted to collect together dispersed matter in the precursors of the bodies.

Under the influence of turbulence, vortical structures present in the fluid flow

Nathanael-Israel Israel: Who is Told by People That He is the Universe-Origin, Life-Origin & Chemicals-Origin Accurate Decoder and/or fluid layers were progressively squeezed, amassed together, and wound up into bigger bodies as fluid layers were pulled and pushed, while movements were being born. The precursors of some bodies rotated more than others. Gravity was progressively established as the fluid layers of the precursors were amassed under the influence of rotation or rotation-like, spiraling motion. The forces that sheared the fluids and compressed them also acted on the precursors according to the intensity of their rotation and their size and, consequently, affected the magnitude of the gravity of these bodies. In other words, gravity and its precursor allowed the precursors of bodies not to be dispersed, but to form unique and unified bodies. Another way of explaining this complex process is that, as the precursors of the celestial bodies were being split and were moving, a squeezing force started compressing and shaping them at the same time that other internal processes were also gathering together the clusters of rocks, minerals, atoms, subatomic particles, and other smaller matters inside of them. Because gravity itself depends on the radii of celestial bodies and their rotational speeds, its value today could not have been established before the bodies reached their current shapes. In other words, as the precursors of the bodies were getting their shape, size, and speed (including rotational speed), a process of gathering was collecting them together and establishing gravity. Just as all bodies today have had their own precursors, so have all forces and interactions in nature. Similarly, gravity, as of today, had its precursor, which was embedded into how turbulence allowed the bodies to be gathered together as they were being split, as I extensively explained in previous chapters. By the time that the size and movement were set and the celestial bodies were put on their "final trajectory" or orbit, gravity was established.

Without a mechanism to match gravity to the size, movement, and composition of the bodies, the process of gathering the bodies' precursors could have highly compressed some bodies, while others could have been scattered or dispersed. As they moved within the fluid layers of the precursors of the bodies, the vortical structures could have pulled some bodies, just as, even today, when a vortex rotates, it can pull matter toward itself. By the time the celestial bodies were formed, a zone of a seemingly "attractive" field was created around them, and most things that are in that zone can be pulled toward that body. On a small scale, all bodies have a field that was formed to exert a force upon matter to define and maintain the form of these bodies. Hence, even at the atomic and subatomic level, there is a field around particles. It is this field that allowed these particles to exist, meaning that without it, matter would have never existed.

Gravity can also be explained by air, gases, and waves originating from the poles due to the rotation of celestial bodies. In other words, the rotation of celestial bodies, and even of other bodies, can cause the formation of a field around them, whose intensity can depend on the size of the bodies, the rotation speed, and other factors. For instance, when the fluid (e.g., water) in a bowl is rotated, a force is born that causes a conical structure at the surface of the fluid, and fluids (e.g., air) are "attracted" or "pulled" from the outside toward the inside. A similar effect is found with hurricanes and many other moving things.

Gravity is higher at the poles because the force of the air being pulled toward the poles directly hits them, while it is lower at the equator because the equator is farther from the poles. The force of the air descending at the poles is accompanied by winds at the poles, which helps explain why the poles are colder than the equator, although some of them are more in contact with the Sun than the equator. Both the mass and the rotational angular speed play a role; consequently, the way the precursor fluid layers were spiraled played a major role. For instance, Saturn is more oblate because the small density of its precursor had predisposed it to be highly squeezed or compressed at the poles. Hence, its sphere was flattened more than that of Jupiter.

I was inspired in 2013, years before I knew I would be writing a book on the origin of the universe, that gravity is like the leftover of the forces that defined the existence and organization of the celestial bodies. To explain this concept, I drew a comparison from living things. Indeed, every living organism and nonliving thing seeks to protect its domain. Human beings, wild animals, and planets do the same. At the national level, countries and states try to protect their borders and communities. At the individual level, human beings like to hide themselves inside fences. In each nation, borders exist to delimitate states, cities, counties, etc. Additionally, human beings protect themselves with clothing to withstand adverse weather. Without those strategies, perilous things could have destroyed human beings.

From herbivores to carnivores, wild animals have their own ways of protecting themselves according to their ecological niche. Although carnivores try to dominate herbivores, while bigger animals try to dominate smaller ones, each wild species manages to have a territory to enjoy. Because they cannot move from one place to another on their own, plants adapt to their environment. They also know how to defend their ecological niche and even compete with other plants in their environment. Plants of the same species seem to enjoy living together more than with others from different species.

When a stranger enters someone else's property, they may be attacked and charged with trespassing. Some immigrants are sometimes caught, detained, and deported from their new country of residence if they do not have their immigration papers properly set. When a stranger enters a new environment or niche of a different species, the latter can oppose, attack, destroy, or try to dominate the former. This is not true just for human beings but also for wild animals and plants. These behaviors hide a code or law of defense not only of ecological niches but also of fundamental laws of existence that sustain life amid environmental challenges and maintain global laws at the root of the formation of the universe.

As far as nonliving things are concerned, matter is equipped with an "ability" to defend itself and its environment. Atoms know how to maintain their integrity and not "voluntarily" engage themselves in reactions or interactions that will destroy them. And when a stranger enters their environment, "frustrations" manifest as reactions, including radioactive reactions that can harm nearby living organisms. Likewise, on the scale of celestial bodies (e.g., satellites, planets, and stars), they know how to defend their environment or propriety. When a stranger enters their environment, celestial bodies can attract them and pull them toward their surface using gravity. In other words, gravity is one of the mechanisms that celestial bodies use to maintain and defend their environment or property in the universe. The impartation of energy, motion, and other characteristics of anything that exists in nature was not meant to be lost easily, nor is the force needed to reverse these laws cheap!

Gravity is not the force that explains the gravitation or revolution of the celestial bodies or their attraction toward one another, but it is a force proceeding from the way the precursors of the celestial bodies were amassed and which ensures the existence and "cohesion" of the celestial bodies. It was a major error that some scientists made in thinking that planets attract their satellites, one another, or even the Sun, and vice versa. The misunderstanding of gravity is one of the main reasons for the scientific drifts, such as those in astrophysics and cosmology, which, in many ways, seem to be, or will soon become, a religion where claims are made and accepted without proof, defended by confusing statistical and other mathematical analyses. On the scale of subatomic particles and atoms, the process that formed them and defined their limits also locked them into compartments in such a way that some theorists can mistakenly think that those particles are attracting one another. In other words, just as some theories on gravity make people believe in an "action at a distance" (e.g., hypothetical attraction between a primary body and its secondary bodies), some people mistakenly think that nucleons and electrons are attracting. In my book on the formation of chemical particles, I explained in detail how they were formed.

By the time I reached this point in my investigation, I realized that gravity would never be sufficient to explain the movement and functioning of the celestial bodies and their clusters. If gravity can be reversed for all of the bodies in the universe, matter may not hold together freely, and celestial bodies cannot maintain their state, and in the end, clusters of matter will be spread out in the universe into smaller particles, similar to the first particle in the early universe, indeed. In other words, anything that can reverse the direction of the rotation and revolution of the celestial bodies, meaning causing celestial bodies to rotate and orbit in a direction and sense contrary to what they are as of today, could also reverse gravity. The systems or processes that progressively established gravity were also connected to those that established rotation, revolution, and many other properties of celestial bodies and particles in the universe. But because human beings and the level of their technology are not and will not be able to reverse the course of the movement of the celestial bodies, they can neither change the gravity of the celestial bodies nor change the design, functioning, and the course of the history, story, and destiny of the universe. What is that destiny? I will tell you later.

19.2. Two types of gravity: spiritual and physical

On December 11, 2020, at the end of a stretch of 5 days of prayer and fasting, I was tired and sitting on my couch when it suddenly appeared to me that there are 2 types of gravity: spiritual gravity and physical gravity. Physical gravity is what secular scientists know or think they know, while spiritual gravity is insinuated in spiritual laws that even most believers seem to ignore. Physical gravity is understood as the ability of celestial bodies to attract objects within their gravitational fields, regions around them under their "control". Spiritual gravity is gained when believers grow in observing God's ordinances, enabling God's grace to grant them access to spiritual realities aligned with God's holistic plan. As a born-again believer grows in the word of God and walks righteously with God, his light increases. The bigger his light, the more things he can attract not only from the heavens but also from other systems of Earth. Those who do not believe in God cannot attract heavenly blessings because they are beyond the domain that their spiritual size and influence can reach.

Just as the gravity of a celestial body has limits beyond which it cannot exercise influence, so also the spiritual gravity of human beings and even of other beings and things has limits depending on the state of their spirit or the substance of their existence. For instance, God cannot hear the prayer of a sinner because the sinner has no spiritual light, which can increase his influence to reach the dwelling of God and attract anything besides what God's grace provides on earth. In contrast, God hears and answers believers because of the light in them, which is the word of God. Oftentimes, when believers offer up prayers, cry out to God, and fully trust Him in trouble or in the middle of a battle, God answers and hands their enemies over to them.

Believers do not have the same influence because they lack the same spiritual gravity. Because some believers are endowed with more grace than others, the spiritual blessings and power they can attain are beyond what their personal devotion to God could have brought. In other words, because of His grace, God gives some people more than they "deserve." Just as intermittence exists with celestial bodies, so also do all human beings do not have the same spiritual gravity, influence, power, and even grace. Likewise, unbelievers and evil forces differ in the magnitude of evil they can commit. Satan has the highest evil spiritual gravity, while demons are ranked differently. In contrast, no being has gravity as high as that of God.

On May 2nd, 2017, as I was thinking about the laws of creation, I felt like people would never comprehend gravity if they could not first comprehend how matter was brought into existence by the spoken word of God before being molded into different bodies. I noticed that gravity is just a consequence of the laws that shape bodies into different forms according to their mission, spatial location, and the movement of their precursors! For instance, according to the Bible, Jesus Christ holds everything together: "*By Christ, all things are held together*" (Colossians 1:17).

Jesus Christ is like the glue that holds things together. Have you ever heard someone say that someone is the glue that holds the rest together? The holding together of things can apply to both physical and spiritual gravity. God must have been the source of the force that held everything together, not only at the individual level but also at the global level. Not only does God hold everything together, but He has also calibrated gravity to meet the needs of living things on Earth. For instance, at the microscopic level, atoms are built on a solid foundation. If nucleons, which are at the core of atoms, were not made very solid, they could have been screwed and made the matter they formed collapse. Subsequently, the universe could have been vulnerable to collapse, and life could have been vulnerable. A shaky foundation cannot hold a strong building, regardless of how strong the building is. Unfortunately, people like to neglect their foundation, rushing through its layout and then spending a lot of energy building beautiful stories that cannot stand on the wobbly ground! Then, they tend to get people to praise them or to blame them for the chaos that will follow the collapse of their unthoughtful construction.

It is to avoid creating a fragile world, for God spent His time (eternity) building the physical world on the foundation of eternal or spiritual things, rather than the reverse. It is not by chance that the Bible says that the universe will pass away one day, but the word of God, which is the foundation of the creation, will never pass away. For if God could easily change His word, He would have already destroyed all creatures, for they are based on the utterance of His word. Hence, even the Old Testament, which some people think is old and obsolete, is not. Although some people think that Jesus abolished the Old Testament, He told people in the New Testament that not even an iota would be removed from it. Jesus even told them He came to fulfill the law, not to abolish it. If more believers understood the Law (the Torah), I believe that they would have had a better understanding of the feasts of the Lord and other important times that hold secrets about the formation, functioning, and fate of the universe. The unchanged ability of God and His word is also a statement about the trustworthiness, reliability, consistency, steadfastness, or perfection of the foundation that God had laid for the world that He created.

Likewise, the foundation of celestial bodies and their clusters is based, among other things, on the foundation of their atoms, subatomic particles, and other smaller particles. Scientists are sometimes shocked that the nuclear force holding the nucleons together is much stronger than gravity, which they think holds celestial bodies together. Scientists think gravity is the weakest of the fundamental forces. If gravity were stronger than the nuclear force, the universe might not be standing, for it could have collapsed because its largest building blocks rely on weak foundations. Also, if gravity were stronger than it is now, human beings might have a hard time lifting their legs to walk, and they might be almost glued to the surface of the Earth, like magnets stuck to one another. For instance, some heavy people tend to have a harder time walking for a long distance than lighter people. Some overweight people have problems with their ankles, knees, and backs. In contrast, a much smaller gravity could have damaged such lives, for it may have even predisposed human beings to be moved like leaves blown by the wind. In short, human locomotion and many other human activities could have been problematic if gravity were different than the current norm. A change of gravity could have also affected plants, wild animals, and many other things.

From a supernatural perspective, God's omnipresence is an expression of the highest degree of His gravity or influence. Although God's presence is not physically present everywhere, it can be felt everywhere. In other words, the omnipresence of God is not that He is physically present everywhere, but His presence can be felt everywhere, and He has things in nature that represent Him everywhere. Even the things in nature recognize God for who He is. When Jesus was entering Jerusalem and believers were worshipping the King of kings and laying their cloaks in front of Him as He was riding on the colt, and the Pharisees became upset and told Jesus to rebuke His disciples, Jesus responded to them saying: *"I tell you that if these keep silent, the stones will shout out!"* (Luke 19:40, Tree of Life Version). If rocks will shout out in praise to Him, they must be considering what they know of Him. Likewise, as he was speaking to the Israelites of the things which God had done and they made a covenant to worship Adonai and listen to no one but Him, Joshua placed a stone declaring their covenant and he told them (Joshua 24:27): *"Behold, this stone will be a witness to us. For it has heard all the words of ADONAI which He has spoken to us. So it will be a witness to you, lest you deny your God"* (TLV).

This implies that even rocks are listening and recording spoken events and words, similar to a witness who sees a crime. The mountains also listen to Jesus and know His authority. As Jesus was teaching His disciples about faith, He told them: *"Because you trust so little. Amen, I tell you, if you have faith the size of a mustard seed, you will say to this mountain, 'Move from here to there' and it will move. Nothing will be impossible for you."* (Matthew 17:20, TLV). If the mountain must move, it must know the authority that is instructing it. Likewise, like a preacher once said, "Some places and beings (planets, animals) represent God everywhere and are recording events and activities. Although God is on His throne, His presence can be felt everywhere, just as a telephone tower exerts influence over a certain distance. God can be felt everywhere, but He is not everywhere." For instance, God is not present in unbelievers. Likewise, by making a man in his image and likeness, God has given human beings, to a certain extent, omnipresent ability. Otherwise, God has failed in transferring His likeness to men. Because God never fails, human beings are like small Gods or God-beings, waiting to manifest themselves. The omnipresent ability of men allows them to dream, having their spirit in them, and yet they can be somewhere else. When some people exercise that ability to some extent, what they do in their dreams can also be manifested in the spirit.

Just as God's influence can be felt from His dwelling, and just as human beings try to exercise influence on others over some range of territories (using power and many other means), so also celestial bodies and particles (small and big) in the universe exercise some influence and authority over some space surrounding them. Even wild animals tend to have a domain over which they try to exercise and maintain their influence. The influence of human beings, wild animals, celestial bodies, and particles in nature generally decreases with distance and varies with their nature. Unlike the influence of God, which is infinite, that of men is finite but depends on the territories they cover. Cities, counties, states, territories, nations, continents, organizations, or associations of human beings also exercise different levels of influence that generally depend on their position. The influence of celestial bodies can be felt as gravity, which allows them to try to capture and "attract" other bodies (usually smaller than themselves) that come within their sphere of influence. As for living things, they fight with others to maintain their influence, which constantly changes position because of their movement and/or the movement of things around them. In other words, the search and exercise of power and influence, the negotiations, wars, and other forms of conflicts or peacemaking found with living things are efforts to acquire, defend, and maintain their territories just as gravity for celestial bodies is an expression or a consequence of their ability to defend and maintain their nature. To put it another way, just as living things need a territory or an ecological niche to live and enjoy their attributes or privileges as beings, so also do celestial bodies and even particles (e.g., atoms, subatomic particles, etc.) need a territory to exist and maintain their characteristics that human beings are still struggling to understand as if things that we think have no life are really inert and have no feeling or emotion. Anyway, in other chapters in this book, I elaborated on the emotions of nonliving things.

To make a long story short, everything in the universe is linked to God and can be traced back to Him who created the world. Likewise, everything in the universe was supposed to gravitate around God or depend on Him, who must be at the center and above everything, and who wants to be worshipped for who He is. We were created to serve and worship Him, not the other way around. That is why God will punish those who refuse to obey His laws and put Him at the center of their lives, while He will bless those who believe in Him and live according to His word, which sanctifies and gives them access to everlasting blessings.

19.3. Other fundamental forces in nature

If scientists cannot properly explain gravity, which can be easily measured with tools that can be seen, how can we even trust measurements and theories done on smaller particles like atoms and subatomic particles that nobody has ever seen? The data I analyzed led me to conclude that although many models claim to explain the so-called fundamental forces in nature, I think they missed the point. For if they had properly understood how things in the universe interact, they could have also understood how everything was formed. The radical explanation of the universe's origin (using turbulence) also calls for a new perspective on the fundamental forces and how the universe functions as a whole.

Indeed, as the turbulent prima materia (the initial matter created by God in the universe) was pushed into motion, destabilized, and entered turbulence, different fields could have been birthed and formed across the changing compartments of matter. These fields could have been precursors to interactions that theorists later twisted to invent the fundamental forces. For instance, the processes that compress or squeeze subatomic particles and even atoms may give the impression of forces between them. But in reality, no subatomic particles carry any force, just as the presumed graviton (which is never and will never be found) is not responsible for gravity. In other words, although it can be acceptable to talk about forces inside atoms, I am not convinced that they are mediated by particles such as those postulated by modern science before my discoveries about the formation of the universe, but they are mediated by the processes that determined the way matter and its clusters were formed. This way of thinking aligns with the explanation of gravity, which does not imply that celestial bodies attract one another. For instance, applying my hands to compress or squeeze something does not mean that the constituents of that thing are attracting one another. The interactions between human beings through love and sometimes hate are strong and may be perceived as a force of love or a force of hatred, yet these interactions or related "forces" are not mediated by a particle, even if the settings of the human heart may play a role. In other words, the interactions between two things do not always have to involve particles. There are things mediated by mysterious forces, beings, and unknown entities that human beings can never understand with their mind, particularly when they refuse to believe in God, the Creator. For instance, the Bible teaches that Jesus is the mediator between men and God, yet that mediation is not aided by any particle known to modern science.

Similarly, by the end of the formation of the universe and its bodies, natural "forces" may seem present between the bodies, but they are not the root cause of the formation of the universe. Some of the "forces" that acted on the precursors of the bodies and that are still acting on their daughter bodies today were produced by their movement, including rotation and revolution. For instance, as bodies rotate, they can attract some air toward their poles. Consequently, the rotation of the celestial bodies can be responsible for the influx of cold air toward their poles. The forces accompanying the incoming air at the poles also explain why gravity can be higher at the poles and why the poles are flatter than the equator, while the equator is oblate.

Unlike what some people think, the so-called fundamental forces were not forged in one step or instantly. However, as the structures and characteristics of the bodies' precursors (on different scales) in the early universe changed, so did the intensity and nature of the interactions between them, some of which would later be called fundamental forces. The characteristics of these forces and the things that mediate them evolved until a kind of equilibrium was reached, and most of the constituents of matter (small and big) were locked into systems whose dynamism is much smaller than what prevailed at the beginning of the world. In other words, the changes that occurred inside bodies or systems of bodies in the universe today are much smaller than those that occurred during the formation of the universe. Some of the current modifications occurring within bodies and their systems are not very significant and may not even be perceived, of course, but many things in the universe are changing. The magnitude of those changes may depend on the size of the bodies bearing them and their localization. Because I did not mean for this book to be filled with too many technical scientific terms, I will have to stop the demonstration of the fundamental forces in nature, hoping that those who want to know can consult my books *"Turbulent Origin of Chemical Particles"* as well as *"Turbulent Origin of the Universe"*, in which I expounded on the origin of the forces in nature and the interactions between them.

'Science180 Academy' Success Strategy:
SCIENCE180 MODELS OF THE ORIGIN OF THE UNIVERSE AND ITS CONTENT

Science180 Models consist of all the theories elaborated by Nathanael-Israel Israel regarding his groundbreaking discovery on the origin of the universe and its content including all forms of life and chemical particles. To learn more, visit Israel120.com/contact. These theories are detailed in various books written by Dr. Nathanael-Israel Israel and encompass the following:

1. *SCIENCE180 MODEL OF COSMOLOGY*, also called Science180 Cosmology, Science180 Model of Cosmology, Science180 Cosmological Model, a scientific theory that explains Science180 to the scientists. Discover the details of this model in Nathanael-Israel Israel's book titled *"Turbulent Origin of the Universe."* In that book, you will also unearth the new physics that will revolutionize science forever and land you in a zone of original ideas that improve lives nonstop regardless of your expertise.

2. *SCIENCE180 CREATIONISM*, also called the Science180 Model of the Creation of the Universe and Life by God, a scientific theory that presents the origin of the universe in a biblical language. If you want to learn more about how to scientifically prove the Biblical account of the creation of the universe and the existence of God in a way that makes the head of God deniers to spin faster than a DJ's turntable, then get Nathanael-Israel Israel's book titled "*Reconciling Science and Creation Accurately.*"

3. *SCIENCE180 MODEL FOR THE GENERAL PUBLIC* (which explains the origin of the universe and life to the general public in a language that laypeople can understand). Find out more in Nathanael-Israel Israel's book called "*From Science to Bible's Conclusions*", a scientifically verifiable, bestselling book to finally get the accurate, jaw-dropping answer that has been rationally shaking believers, skeptics, and freethinkers. Get this very popular book today.

4. *SCIENCE180 MODEL OF LIFE-ORIGIN*, or the Science180 Model of the Origin of Life, a scientific theory that explains the origin of all forms of life using turbulence. To unlock the step-by-step pathway to decode the origin of life and get the power, freedom, and boldness to detect, correct, and remove all misinformation, ambiguity, and misleading claims and theories surrounding the origin of life and take advantage of the opportunities that an accurate understanding of the life-origin creates, get Nathanael-Israel Israel's book titled "*Turbulent Origin of Life.*"

5. *SCIENCE180 MODEL FOR CHILDREN*, a children's version of the theory of the origin of the universe and life in a language that 7-12 -year-old children can properly understand. To know the proven formula that helps children to easily answer their huge universe-origin and life-origin questions with confidence, humor, and joy, get "*How Baby Universe Was Born,*" the pragmatic book that has been causing children to belly laugh and thank those who offered it to them.

6. *SCIENCE180 MODEL OF THE ORIGIN OF CHEMICAL PARTICLES*, a scientific theory that explains the origin of chemical particles with the perspective of Science180 Turbulence. If you want to professionally learn how to transform the true knowledge of the origin of chemical particles into insights that significantly add value to your life in less time and successfully establish you as a symbol of freedom, power, creativity, and originality in your field of expertise, get Nathanael-Israel Israel's book "*Turbulent Origin of Chemical Particles*", THE ultimate how-to guide for great people wanting to correctly decode the origin of the chemicals and positively transform their lives. Get this celebrated book today. Don't wait!

7. SCIENCE180 MODEL OF PSEUDEPIGRAPHA, a deep explanation of the secrets of the origin of the universe and life revealed a long time ago, but hidden from the general public. To discover how the only ancient blueprint has the reliable power to help you to accurately decrypt the spiritual origin and history of everything in the universe, get Nathanael-Israel Israel's book called *"Origin of the Spiritual World"*. In it, you will discover deep rejected secrets that have prevented humankind from unearthing the beginning of the universe and know how to properly use the lost and rejected scriptures to articulate the process by which the universe was formed, so you can use that insight to improve your understanding of the Bible, innovate in your domain of interest, and improve your life perpetually.

8. SCIENCE180 MODEL OF THE PROOF OF THE EXISTENCE OF GOD, a theory that ties together most of Nathanael-Israel Israel's discoveries that scientifically prove the existence of God. With Nathanael-Israel Israel's book *"Science180 Accurate Scientific Proof of God"*, you will surely know the only way to scientifically know if God exists and, if so, which of the thousands of beings worshipped across the globe is the true God. In that book, you will also discover the errors in the scientific and religious theories (about the origin of the universe, life, and chemicals) that are putting you at a high risk you will never recover from if you don't quickly and confidently learn how to rationally take control over threats lurking at the edge of your efforts to understand the universe and life today.

9. SCIENCE180 THEORY OF EVERYTHING, (also called the theory of all theories), ties together everything in the universe into a single theory. Checkout Science180.com to learn more about the incoming book that covers this extremely important topic.

CHAPTER 20

CAN YOU BE REALLY FREE FROM EVERY DOUBT ABOUT THE FORMATION, MISSION, ACTIVITIES, AND CHARACTERISTICS OF ANGELS ... (WHAT ABOUT THE UFOS, DEMONS, AND THE OTHER ANGELIC BEINGS—OR DO YOU THINK THEY ARE FAKE NEWS?)

I could have talked about the formation of angels long ago, but because some Judeo-Christian references suggest that the precursors of celestial bodies underwent changes before angels were formed, I decided to explain the formation of celestial bodies before addressing the formation of angels. Because angels are beings that require a location or habitat to live in and perform their duties or accomplish their missions, it also sounded good to me that God could have first created a place to host the angels before creating them. Just as plants, wild animals, and human beings living on Earth were formed after the formation of Earth (their home) was completed, so also angels could have been formed after the creation of the places they would call their primary home. As of today, angels can move across many habitats, but they still have their own dwellings according to their ranks, for not all angels live in the presence of God, nor are all angels the same. And to be exact, just as living organisms on Earth were formed using materials (e.g., dust, soil, and water) from the Earth, so also angels were formed, not created out of nothing, but using the matter that is from the places they would live. God carved angels using fire. Although you may not believe it, almost everything in the universe is controlled by angels. Some angels just sing "holy, holy, holy"; others control the spirit of fire, wind, clouds, darkness, hail, thunder, lightning, cold, heat, winter, spring, fall, summer, etc. As you read the upcoming segments, you will better understand what I mean!

20.1. Who are angels, and why, when, and how were they formed?

God created and formed many things and beings in the spiritual world, such as angels, angelic weapons, garments, tools, and other things that a mere human, limited by his own spectrum, cannot see. Although the Bible says a lot about how human beings, wild animals, and plants were formed, it says little about the origin of angels. The Bible provides ample information about the mission of angels, at least as it pertains to human beings. This implies that it will be difficult to address the formation of angels without referring to other Jewish books not found in the Bible or to other sources of information, such as prophecy. Before and after the Bible was compiled, God had revealed certain things to believers, including prophets, that are not mentioned in the Bible. In other words, it is important to be open to the fact that not everything about angels is revealed in the Bible. I have heard believers recounting their encounters with angels in a language not found in the Bible, yet they are authentic stories. Likewise, the Books of Enoch and the Book of Jubilees (both of which are quoted in the Bible) provide significant information about the origin of angels. Therefore, to avoid confusing some people, I will not delve into many details about the formation of angels here, but I will use only biblical references to illustrate how their characteristics fit with how they were formed.

In this chapter, I also address the spirituality of physical things, which some people think contain nothing spiritual. There is a high-ranking hierarchy among angels. In fact, during my research on angels, one of the things that impressed me most was how highly organized they are. The third book of Enoch (the Hebrew Enoch) details the organization of angels, from the lowest to the highest, which are closest to God.

God created spiritual things for His own pleasure or comfort, to serve Him and the human beings He created after them. For example, angels carry God's messages to human beings and represent God before humankind. Many people have encountered holy angels, believing they were dealing with God. Angels are very energetic and powerful spiritual beings, yet there are things that human beings know but that angels ignore. Angels move very fast and can accomplish tasks that are otherwise impossible for human beings. Some angels carry the glory of God and can shine many times brighter than the Sun. They have different ranks, levels of authority, and levels of power. Those called archangels are usually considered to be more powerful than the others.

Because of the nature of the material they are made of, angels can stay in hot conditions, such as in the Sun. In the apocalyptic story, Apostle John revealed that he has seen an angel in the Sun: "*And I [John] saw an angel standing in the sun; and he cried with a loud voice, saying to all the fowls that fly in the midst of heaven, Come and gather yourselves together unto the supper of the great God*" (Revelation 19:17). Some angels can appear like stars, and others have the appearance of a star just as it was explained to Apostle John: "*The mystery of the seven stars which thou sawest in my right hand, and the seven golden candlesticks. The seven stars are the angels of the seven churches ...*" (Revelation 1:20).

According to the Book of Enoch and the Book of Jubilees, angels were formed on the first day of creation. According to the calculations I did, as the story of Genesis mentions, the formation of many bodies in the universe was completed within days after the beginning of creation, meaning that by the time angels were created, God had already created other things. This implies that angels do not know all of God's creative work. Unless God revealed it to them, some angels could be ignoring many details about creation. Angels were the first beings God created, meaning they existed before plants, wild animals, and human beings. In Job 38:7, God informed Job that the angels were glad when they saw the things that God created, suggesting that, at least, the Bible confirmed that angels were formed before the formation of the Earth and its hosts:

> "*4 Where were you when I [God] laid the foundations of the earth? Tell Me, if you have understanding. 5 Who set its dimensions—if you know—or who stretched a line over it? 6 On what were its foundations set, or who laid its cornerstone—when the morning stars sang together, and all the sons of God shouted for joy*" (Job 38:4-7).

This also implies that the report in the book of Job about the joy and amazement that angels felt when they saw the creation of God could not have been about their witnessing of everything, but rather about things whose formation was completed after the formation of those angels. God first focused on the creation and formation of spiritual things and spiritual beings, meaning that spiritual things are a little older than physical ones known to most human beings.

Although some people may not accept or believe that angels were formed on Day 1, the ability of angels to change their form and take other shapes suggests to me that their constitutive matter resembled the turbulent prima materia (i.e., the initial matter that God created), and which was transformed into other matter and bodies in the universe. In other words, the ability of angels to transform into other forms is a consequence of their formation during an early stage of creation and the ability of the turbulent prima materia to become anything. I am not saying that angels are made from the turbulent prima materia, but from a materia near it.

Many angels are reported, even in the Bible, to have a fiery aspect, suggesting that they are made of fire. Just as some celestial bodies do not express fire on their surfaces, some angels also do not appear like fire, although they can have a fire nature within, and their outer bodies could be transformed to take other forms.

Considering certain stories I learned about angels, God did not tell angels how they were formed. Until today, I would not be surprised that many angels still do not know how God created them or men. For unless God reveals it to them, they cannot dig into certain secrets of the universe. At one point, God revealed some of those secrets to certain angels, who then transmitted them to human beings. Yet many angels marvel at how much human beings know compared to them. Moreover, many holy angels know a lot more than most human beings, for all angels are not ignorant of heavenly secrets.

20.2. Desire, will, and disobedience of angels

All angels were created holy, but at one point, some spiritual beings who had once obeyed God disobeyed Him. The master of these disobedient angels is Satan, formerly known as Lucifer. The angels that disobeyed God are generally called demons. In other words, Satan and demons are all fallen angels, meaning they sinned against God and consequently, their initial state was changed or denaturalized. Knowing that some angels disobeyed God while others obeyed Him, I think that angels have emotions and a will, and are not always controlled by God or the Holy Spirit. God did not create angels to be like robots who would always do His will. Otherwise, some angels would not have fallen. However, the angels who remain obedient to God are ministering spirits, some of whom bear the presence of God.

As I studied the ranking and organization of angels, I understood that some of them could never have access to the throne of God, while others are near it all the time, singing and worshipping the Holy God: "Holy, Holy, Holy is the Lord …". As of today, when angels are mentioned, people tend to think of those who still obey God, while when demons are mentioned, people think of fallen angels. God allows some angels to stay on Earth and work with and for human beings, under the control of higher-ranking angels and some human beings who are "highly" anointed to interact with them. In other words, some human beings can directly interact with certain angels and instruct them to do specific things. I understand that some angels are directly influenced by God's thought, while others (e.g., messenger angels) are influenced by the thought of the angels they follow. That is why some angels followed Satan and they fell. Obedient angels (or holy angels) do not follow things outside of the will or plan of God. Consequently, if human beings give them instructions that are not aligned with the will of God, holy angels will not execute them. That is why some people understand certain things through the word of knowledge and then prophesy things that never happen because the intention attached to those prophecies is not aligned with the will and timing of God so that holy angels can execute them. Similarly, evil people can give instructions to demons and they execute them to the extent of their ability. For all demons are not ranked the same.

Angels can transform themselves into other things. Just as in the early developmental stages of an embryo, the initial cells can be changed into any type of cell (an ability termed "totipotency"), so also angels and any material resembling the turbulent prima materia can easily change into anything. In contrast, human beings are not able to transform into other things because their bodies cannot easily be changed; yet, some contortionists can apply some extraordinary shapes to their body. Some anointed men of God have also testified to have passed through walls as their bodies took another form or operate from another dimension. Some witches can also take different forms to some extent. Some wild animals, such as chameleons, can mimic their environment, while others can even metamorphose (caterpillars also turn into butterflies) into others at least during their developmental stages, suggesting that, to some extent, it is possible for some living organisms to be transformed into other things.

Some people can argue that angels are asexual, but if that was the case, why are some angels reported in Genesis 6 and elsewhere to have slept with women? The obedience of angels which caused some not to marry, may not be sufficient to attest that all angels have no sexual organs. Likewise, Jesus' assertion that, in heaven people will not get married may not be sufficient to prove that angels are asexual. Even if angels have no sex, their ability to transform themselves into other forms of beings could have allowed some angels to make for themselves sex and then defile themselves by sleeping with women. But the main point is that angels are not supposed to have intercourse with women.

Before his fall, Lucifer was a cherub, meaning one of the chief angels holding one of the highest ranks. As Lucifer was rebelling against God, many other angels, including many who were under his authority did the same and followed Satan in his fall, and God removed them all from His presence. In the Book of Revelation, Apostle John revealed that 1/3 of the angels in heaven followed Satan in his fall. The initial dwelling of Satan was not Earth, but another place higher, for his fall was described as a descending movement.

Based on Biblical stories, people may think that Satan fell after the creation was finished. For the Bible talks about Satan trading in the Garden of Eden before his fall, implying that he could not have fallen before the Garden of Eden (mentioned on Day 6) was planted for Adam to dwell in. Even if some people consider the Garden of Eden to have been implanted by God rather than formed directly on Earth, it is important to know that God created the Garden of Eden as well, even if some believe it was imported from elsewhere. And its creation must have happened before the 6th day of creation, when Adam was formed.

It is unfortunate that some people believe in and defend theories that assume that Satan fell on the first day of creation in Genesis 1:2, where the earth is said to have been chaotic. Although some Jewish writings suggest that Satan tempted Adam and Eve the same day they were formed, meaning the 6th day of creation, other Jewish books suggest that the fall of Adam and Eve could have happened at least 7 years after their creation. I provided more details in *"Origin of the Spiritual World"*.

Before the fall of Adam and Eve, the control of the Earth was in the hands of Adam. After his fall, Satan took over the control of the Earth. Since then, Satan has been trying to cause all living things to disobey God and His commandments. Consequently, the equilibrium between the universe and its hosts has been broken since the fall of Satan, Adam, and Eve. Because God has hidden some natural laws, neither Satan and the demons, nor human beings have been able to understand all laws sustaining the universe. Otherwise, they could have totally destroyed what God had created. Therefore, until the end of the world, evil will continue to dominate on Earth, for the current age is dominated by Satan and demons. Yet, in the end, God will prevail!

20.3. Consequences of the fall of Satan and his angels

Fallen angels had to be removed from their initial state, for their corrupted nature is incompatible with their original state. Hence, fallen angels lost their brightness, glory, or light, but are filled with darkness, although they can pretend or act as angels of light. That is why Lucifer, who was an angel of light before his fall, is now a dark angel, for light is no longer found in him. Consequently, because they are denuded of light, fallen angels like darkness but hate light. For light is incompatible with the mission they have taken upon themselves after their fall. Fallen angels have positioned themselves in different spheres of life on Earth, not only in the physical darkness of space, but also in the spiritual darkness of living things.

Although conscious of their state, fallen angels are not as knowledgeable as human beings, particularly believers, who possess mysteries about God's plan. Some angels, even those who are still obedient to God, are very powerful, but still, they do not know as much as human beings. For in the beginning, human beings were created superior to angels and were even able to directly communicate with them and instruct them. But since the fall of Adam and Eve, not only was the communication between men and angels broken or reduced, but they could no longer receive commandments from men. For the nature of human beings has been corrupted, and holy angels can no longer submit to them without always ensuring that they themselves will not be in trouble.

During my studies, I came across a book called the "First Book of Adam and Eve," which presents some shocking accounts of the angelic nature of Adam and Eve before their fall and the loss of their privileges afterward. When I read that book, which, according to the Book of Enoch, was written by Enoch under God's commandment, I was shocked by how many privileges sin has caused humankind to lose. Yet, human beings are still sinning and complicating their existence on this Earth, where they were supposed to rule over everything. Because human beings cannot easily see them with their physical eyes and because of the struggle for power, fallen angels manipulate human beings, blind them, and cause them to do things that, in the end, are meant to destroy human beings. At the same time, some believers whom God has given the grace to recover some of those privileges act as if it were possible for them to dominate everything again in this world before the reign of Jesus in the millennium kingdom, which I will address toward the end of this book.

Since their fall, Satan and demons hide themselves in all sorts of darkness and hurt human beings. Beforehand, there was no evil in the darkness. Since Satan and demons turned darkness into evil, it is not surprising that they will be in darkness in eternity. The corrupted power of evil spirits is not enlightened by the real knowledge that could have enabled them to achieve all of their wicked goals. Therefore, they are limited in what they can achieve, causing God's universal plan to continue its course despite Satan, demons, and some human beings trying to stop it. The eternal condemnation of evil spirits is also due to their preference for darkness and the null probability that they will never come to light to receive enlightenment and freedom. In other words, fallen angels are already condemned, and there is nothing anyone can do to change their sentence. Subsequently, they persist in their state, imagination, and acts, which cause them to look after things that they can never attain but that keep distancing them from God forever.

Likewise, as time passes by, some unbelievers harden their hearts and persist in their dark behaviors, not knowing that they are working out their eternal destiny in hell. Ignorance of spiritual things and even of physical things far beyond current human reach is one of the things that keeps people in the darkness of celestial realities. For the dwelling of God is filled with light, and no one filled with darkness can access God, nor His treasures. Consequently, those who fill themselves with the word of God, which David describes as a lamp, enlighten themselves to receive from God and be guided by His instructions. Every time a Christian reads and applies the Bible, he absorbs lighter and is more qualified to be in God's presence one day.

In the end, those who refuse to believe in the light of God will spend eternity in the darkness reserved for Satan and his demons, while those who believe in God are led to the eternal, enlightened dwelling of God. In other words, the lives people are living today on Earth are a preparation for their eternal experiences. Even if Satan and his demons were offered Paradise, they would not want to go, for their love for darkness and their ego or pride would not let them get close to any light. Hence, by their own nature, fallen angels have eliminated themselves from ever living in heaven. In contrast, despite the pressure that Satan and demons put on believers to cause them to embrace dark choices that will land them in hell, believers do not yield to that pressure, therefore rejecting hell, while treasuring heavenly things in their hearts. The stiffness and hardening of the heart of Satan and demons have caused them to persist in the negative thoughts that they may still believe to be true. Some of the spiritual beings that God created fell and have been rebelling against God and other creatures. Apostle Paul warned his followers about those beings: "*For we wrestle not against flesh and blood, but against principalities, against powers, against the rulers of the darkness of this world, against spiritual wickedness in high places*" (Ephesians 6:12).

Just as God cannot force anyone to go to heaven, He can neither force anyone not to go to hell. Hence, before the beginning, God planned heaven and hell long before creating anything. In other words, God was not surprised by the rebellion of some angels, nor by the obedience of believers. The darkness of hell would be a perfect retribution for those who refused to believe in God, while the treasure in heaven is also a perfect reward for those who chose to believe in God during their life on Earth.

The spirit of God, which, as of today, lives in believers, enlightens their paths and guides them on their earthly journey to heaven, revealing to them secrets that unbelievers may never accept or comprehend without the faith they rejected. In contrast, because they lack the true light, which is Jesus (John 1:9), unbelievers cannot see or understand eternal things, hence they reject God and live as if there is no life after death, until they die before realizing that, in fact, there is life after death and their life after death is denuded of the author of life, hence an eternal death filled with punishment for their unbelief while on Earth.

Unaware of these facts and other spiritual laws hidden behind physical things, many people uselessly wage war against things that do not reflect the reality of the duties capable of preparing them for eternity. Those who understand those duties are usually attacked by those who are blinded, hence in most human societies, the liberators are not welcomed, for what maintains most human beings in suffering and oppression is not physical, but spiritual. This can also explain why some of the ignorant of spiritual things are those who are highly educated in secular schools, where realities about God are rejected in favor of useless theories, which have nothing to do with the truth. Yet, those theories are usually applauded by crowds that do not know where they are going. Their defenders ignore that the greatest knowledge is spiritual, not physical. Physical knowledge is limited, while spiritual knowledge can benefit its owner today and forever.

Those who believe in God rest in peace even in the midst of their earthly troubles, often caused by demonic and jealous forces. Unbelievers reject all spiritual things accepted by faith and focus on physical things, which cannot be understood without the underlying realities grounded in the spiritual. Unbelievers often want to see before they believe. Yet the Bible says, "*So faith comes by hearing, and hearing by the word of the Messiah.*" In Hebrews 11:1, the Bible says, "*Now faith is the substance of things hoped for, the evidence of realities not seen.*" Hence, a great blessing is associated with believing in the truth without seeing it (John 20:29). For even most who believe in God cannot always demonstrate their belief through physics, but rather through faith again!

As of today, most physical things host gigantic spiritual forces that most human beings cannot see. Yet some wild animals can even see better than human beings. In other words, some wild animals see and perceive spiritual things that human beings cannot see. Unfortunately, those wild animals are unable to use their knowledge to act in ways that add value to themselves, as humans try to do.

All forms of hypocrisy, lies, sabotage, theft, violence, unbelief, immorality, and other sexual perversion are manifestations of dark behaviors. And because they ignore their true enemies, most unbelievers delight in working for their own enemies. The division and struggles between local and international agglomerations of human beings and nations illustrate the spiritual struggles hidden behind physical phenotypes of discomfort. As I finished working on this segment on August 18, 2016, I praised God for the grace He gave me to tap into some of the hidden mysteries of the eternal!

20.4. Some UFOs are angels and demons

As I deal with angels and demons, I felt like it would be appropriate to deal with UFOs here. Although their characteristics, mode of action, and origin may still not be understood by government officials (who seem to be the most interested in understanding them, though they could also be using some of them as military tools), I think some UFOs are angels, including demons and holy angels. In other words, not all UFOs would be considered evil; some may be holy angels on a mission on Earth that human equipment can detect.

Considering that demons can be a type of UFO, our world is already invaded by UFOs. Similarly, if angels are considered a form of UFO, the Earth will be more frequently invaded by these types of UFOs. As of now, some angels interact with people on a daily basis; some people may realize they are dealing with angels, while others may have no clue they have encountered them. In other words, some people may have met angels without realizing it, for sometimes angels can behave like ordinary human beings, while at other times they can look very strange (some have many faces, many eyes, many wings, etc.). In fact, the Bible says, *"Do not neglect to show hospitality to strangers—for in doing so, some have entertained angels without knowing it,"* (Hebrews 13:2).

Moreover, the Bible says that in the last day, God will go against the Devil and his demons in a huge war called the battle of Armageddon (Har-Megiddo or Mount Megiddo). In Revelation 16:16, it is said, *"Then the spirits gathered the kings to the place called in Hebrew Har-Megiddo."* This war will be brought on by the angels of God, who will go against some unbelievers who are fighting God's people. According to the Bible, the angels of God will win the battle, and this victory will precede the coming back of Jesus Christ. Many other religious people believe that, sooner or later, this world will end with the culmination of angelic activities. Therefore, one day, human civilization will be invaded by other beings that scientists can also call UFOs. Nevertheless, it is important to notice that many people believe that certain angels of God and God Himself are more powerful and perfect than any mere UFO that human beings can imagine.

Although they do not fully understand what they are talking about when they refer to UFOs, and although some of their predictions are imaginary, some scientists are predicting things that are not completely false: an invasion of our societies by UFOs. What some scientists do not know is that nothing they can do can prevent this invasion or the unbelievers from being defeated during it. There is no bomb that scientists can make to prevent the angels of God from fulfilling God's commandment by warring against the Devil, catching him, and putting him in a prison for 1000 years before later ending this world and sending the unbelievers to hell, while God will take the believers to heaven where they will be with Him forever and ever! My friend, are you ready to meet your Creator, the only God?

20.5. Everything has a spiritual component

Everything in the universe has a spiritual component. Although most people think that spiritual things cannot be seen, it is important to know that human beings cannot see certain spiritual things because of their limited spectrum. The spiritual world is a tangible world, which can be felt, touched, and seen. Unlike what some people think, many things that seem to lack life can express emotions and movement similar to those of living things. For instance, David, known as the man after God's heart, knew that even celestial bodies can speak and worship God: "*Praise ye him, sun and moon: praise him, all ye stars of light*" (Psalm 148:3). David also suggested that even waters have feelings. If not, why would he exclaim the following concerning the emotion of waters: "*The waters saw thee, O God, the waters saw thee; they were afraid: the depths also were troubled*" (Psalm 77:16).

If water cannot listen, why did the sea calm down as Jesus rebuked them as if they were trying to sink the boat in which Jesus and his disciples were? Do I need to recall that even the Red Sea split at the order of Moses when the Israelites were leaving Egypt and returning to Israel? Did you ever wonder why living and nonliving things respond to prayer and obey some orders given to them?

Everything God created can listen to and respond to stimuli, including the voices of not only believers but also unbelievers. Matthew 17:20 says that believers can tell a mountain to move, and it will move.

The physical world that most people see is a compressed or clustered version of the spiritual world, just as matter in a gas state can be compressed into a liquid or solid state that can be touched. For instance, although it is impossible for the naked eye to see water molecules in the air, when water molecules condense or freeze into ice, even the profane who never went to school can see and feel it. The same is true of many spiritual things that most people cannot see because they are in an expanded form, visible only by a grace given to a few. Hence, it is a mistake to think that physical things are not spiritual. Even the Earth we live on can be reversed back into an original precursor state that the human eye cannot see, just as a powder can be released into the air and its molecules can become invisible because they are spread over a large portion of space! A few centuries ago, long before the discovery of microscopes and telescopes, for instance, no human being could have believed in the existence of particles, organelles, or deep details known today in biology, chemistry, and space science. In other words, over time, some things perceived today as spiritual can be easily seen in the physical as more advanced tools become available and understanding deepens.

The Earth can hear: "*Come near, ye nations, to hear; and hearken, ye people: let the earth hear, and all that is therein; the world, and all things that come forth of it*" (Isaiah 34:1). If the earth and heavens could not speak, why would God say the following? "*And it shall* come to pass in that day, I will hear, saith the Lord, I will hear the heavens, and they shall hear the earth" (Hosea 2:21). If the weather cannot hear and respond to prayer, why did Elijah pray, and it did not rain for years? "Elijah was a man subject to like passions as we are, and he prayed earnestly that it might not rain: and it rained not on the earth by the space of three years and six months. And he prayed again, and the heaven gave rain, and the earth brought forth her fruit" (James 5:17-18).

Joshua commanded the Sun and the Moon, and they stood still just so that the Israelites could win a battle against the Amorites (Joshua 10:12-13), thereby showing that even celestial bodies can listen to and respond to commands. Otherwise, why did the Sun and the Moon stop moving when Joshua spoke? "Then spake Joshua to the Lord in the day when the Lord delivered up the Amorites before the children of Israel, and he said in the sight of Israel, Sun, stand thou still upon Gibeon; and thou, Moon, in the valley of Ajalon. And the sun stood still, and the moon stayed, until the people had avenged themselves upon their enemies. Is not this written in the book of Jasher? So the sun stood still in the midst of heaven and hasted not to go down about a whole day" (Joshua 10:12-13).

Even the Sun knows its mission: "He [God] appointed the moon for seasons: the sun knoweth his going down" (Psalm 104:19). The psalmist could say the following: "Praise ye him, sun and moon: praise him, all ye stars of light" (Psalm 148:3). If celestial bodies such as the Sun and the Moon do not have feelings, why would Prophet Isaiah say the following of their emotion during the millennium kingdom? "Then the moon shall be confounded, and the sun ashamed, when the Lord of hosts shall reign in mount Zion, and in Jerusalem, and before his ancients gloriously" (Isaiah 24:23). Being of a different kind, celestial bodies should not be expected to have the same state of soul as living organisms.

Fire can also listen and respond to commands. Many times, during his ministry, Prophet Elijah called for fire, and it appeared, making him the "only" prophet in the Bible known as the prophet of fire. For many of his prophetic deeds involved fire, including him calling fire from heaven to devour some people who were disrespecting him: "And Elijah answered and said unto them, If I be a man of God, let fire come down from heaven, and consume thee and thy fifty. And the fire of God came down from heaven and consumed him and his fifty. Behold, there came fire down from heaven, and burnt up the two captains of the former fifties with their fifties: therefore, let my life now be precious in thy sight" (2 Kings 1:11-14).

Like Apostle Paul said, everything God created knows and understands God and His power: "For the invisible things of him from the creation of the world are clearly seen, being understood by the things that are made, even his eternal power and Godhead; so that they are without excuse" (Romans 1:20). This implies that people are without excuse, for a creature would not say that it does not know God. Additionally, the whole creation can feel pain and is in fact in pain: "For we know that the whole creation groaneth and travaileth in pain together until now" (Romans 8:22).

Among living organisms, ears are not the only organs that can hear. The eyes are not the only organs that can see. Every human organ can hear and speak, for human bodies are spiritual. As a contemporary man of God said, the kidney is one of the most powerful organs for both physical and spiritual purification. A kidney can also store information and instruct people in a dream. The blood can also hear, speak, and store information. The Bible says that life is in the blood and that the blood of Jesus gives us life. Curses spoken on people are stored in their bodies and activated in due time, sometimes with the help of demons. Similarly, blessings can be stored and activated at the right time. Prayers can remove curses, and spiritual kidneys can filter curses and harmful programs from people's bodies, just as a physical kidney can remove waste and toxins. Hence, cursed people can be delivered, while others can be cursed. When curses are not removed from the blood, they can be transmitted to many generations. In the same manner, some blessings are generational, and no curse can break or remove them. For instance, the blessing of Abraham and his (spiritual) children cannot be cursed or affected by anything. That blessing can retaliate and send any curse back to the sender. Without the ability of human organs, including blood, to hear, store, transmit, and respond to information that can be deposited, curses, blessings, deliverance, and other spiritual operations done on human beings would be impossible. Also, the spiritual bodies of human beings are involved in, fought, and won many battles that the physical bodies and the consciousness of human beings are not always aware of. In other words, not everything that the spiritual being does is transmitted or reported to the physical being. The spiritual part of human beings does not always inform the flesh of physical bodies to know everything. But some people are so established in the spiritual world that they can know exactly what is happening in both the spiritual and physical worlds. Those who are limited in their understanding of the spiritual world are limited in their interpretation of physical things around them.

Because they know deep things that unbelievers do not and cannot know, believers in God are altogether a secret society, which unfortunately often ignores its power. In other words, believers in God are a secret society that those who do not believe in God cannot decrypt. Regrettably, many believers do not know their power, ability, and level of influence, just as the unbelievers do not accept or believe in them. Those who are spiritual know this dichotomy, for they understand the physical better. Hence, the devil knows more than some born -again people know, while some supernatural born-again Christians do not know much about the physics seen from the perspective of the secular realm of knowledge. Hence, some very spiritual people make physical or scientific mistakes that some perceive as basic, while many scientists do not know much about spiritual reality. Likewise, those who fight all battles in the physical realms are missing much, for they neglect the reality, substance, or meaning of their pursuit.

Problems and matter can be formed from spiritual beings and things. Some physical things we encounter in life are spiritual realities transformed into physical ones. For instance, some problems are spirits that transformed themselves into problems and even into living and nonliving things. The Book of Jasher talks about Abraham's dealings with a flood on his way to sacrifice Isaac, but that flood was the Devil, who transformed himself. After Abraham and his child managed to pass that flooded street, the flood disappeared as the Devil failed to drown them! In the Old Testament, Jesus transformed himself into a Rock and living water that followed the Israelites throughout their journey in the wilderness, providing them with water. The angel of the Lord appeared to Moses in a bush that burned with fire. The donkey of Balaam was blocked by an angel, which the prophet riding it could not initially detect.

Considering what Jesus told His disciples in John 6:60-64, the words of God are spirit, and they give life. Therefore, because the whole world was created with the word of God, there is the Spirit of God, or at least a kind of spirit in them, and the word contains life or at least a form of life. Likewise, every food contains a form of life; otherwise, it could not feed and give life to living organisms. The ultimate giver of life is Jesus, and the life He gives is eternal, while the life in matter is temporary!

20.6. Technological demons

Although people think about spiritual beings when they refer to demons, it is important to know that, as of today, many technologies are demonic applications of knowledge. I have come to realize that most people and even nations who are well aware of spiritual demons are not much alerted to scientific demons. Some Africans are very experienced with spiritual demons but less aware of technological and scientific demons, while most people in Western countries do not know much about spiritual demons but are very informed and educated about scientific and technological demons in laboratory settings. A lot of technological things are rooted in scientific discoveries or inventions based on demonic things, meaning things that go against divine laws. This also means that many things and products made in the name of technology are just demonic things that are indirectly carrying the agenda of Satan and demons. This does not mean that every technology is demonic. This calls for the importance of spiritual discernment. By lacking the spiritual discernment of things, some scientists have undermined the understanding of physical things. Everything has a spirit that calls for spiritual discernment. Hence, no one can ever discern anything properly without discerning the spirit behind it.

As a contemporary man of God put it, *"the physical earth is a department in the spiritual earth, and despite the physical form of the earth, it is a spirit thing. Spiritual things can also be seen, felt, touched, etc."* It is kind of like how various stores in a shopping mall are different department stores, all in the same building. I heard a contemporary minister saying that the majority of the earth is spiritual and the smallest ingredient is physical. Because God formed physical things using spiritual things, the spiritual aspects of things can never be extracted from physical things.

Born-again believers in Yeshua have power over everything God created by controlling the spirits of those things. For instance, poverty is a spirit. Some people tend to celebrate miracles only when the supernatural manifests in the physical, but in fact, many miracles occur in the physical with no apparent connection to the supernatural, and people fail to appreciate them. Not everything physical in the physical earth is physical. Spiritual beings can disguise themselves as physical things that a mere human lacking discernment would not recognize, and after mistreating them, they can suffer consequences. It is therefore important for human beings to discern the spirituality of physical things.

Listening to the teachings of many contemporary men and women of God, including prophets, I felt that God is still hiding many mysteries concerning angels and creation from human beings. However, the Books of Enoch, which many believers unfortunately have come to reject because the forefathers who compiled the Bible made the mistake of removing them, have a lot of deep information on angels. For instance, Enoch also explained how celestial bodies are under the influence of angels. When I first read Enoch's description of the spiritual world, I was shocked, and it took me more than 5 years before I could understand the depth of some of his revelations. In short, I have not read any book more spiritual than the 3 books of Enoch. I highly encourage you to read them, for although you may not understand all of them, you may learn something new. According to Enoch, the spirits of things were created before the things themselves. I personally believe the Books of Enoch were rejected by the ancestors because they could not explain them and because they contain certain statements that contradict the commentaries advocated by the rabbis (teachers of the law). For the complete narrative about the formation, types, description, organization, and mission of angels, please refer to my book *"Origin of the Spiritual World."*

20.7. Formation of life on Earth

The Earth is filled with various creatures whose characteristics and diversity have puzzled those who have tried to decode their origin. Due to the complexity of this task, I devoted a different book to the origin and formation of life. Here, I will elaborate on a few features of the beings on Earth. Indeed, considering what I know about the turbulence that molded the celestial bodies and particles in the universe, and what I know about life and the study of living things, it appeared to me on December 8th, 2020, that a highly coded program of biological turbulence was at the root of the formation and diversity of living things. To put it another way, just as an abiotic (meaning not related to, or not caused by, or not derived from living organisms) program of turbulence was run through the turbulent prima materia to produce many precursors of bodies, which in the end were split-gathered into various celestial bodies and particles, so also on the scale of the living things, a biological program of turbulence was run by God throughout the waters (precursors of aquatic beings), and throughout the soil or dust of the earth (precursor of terrestrial living things), and the precursors of other beings (e.g. angelic beings) so that those precursors could be split-gathered into different living organisms according to their order, rank, class, and mission. Just as for celestial bodies, an intermittence (e.g., of size and energy) existed and was the cause of the formation of small bodies between bigger ones, so also for the living organisms, small organisms were formed between large ones. Hence, in the same field, giant organisms like elephants emerged, while organisms smaller than ants and even microorganisms also emerged. In the waters, in the air, in the earth, and on the earth, various organisms were formed. Human beings were the only organisms that God made with His own hands in the image and likeness of God Himself!

Just as a computer program can be written, and after launching it, it runs by itself and executes all the code involved, the biological systems and beings in the universe are running under the influence of turbulent programs coded by God during the creation, some of which are functioning under programs of reflexes, while others are influenced by the will of humans and demons, but everything is under the control of God's master plan. In other words, the whole universe was formed under the control of a master program that can never be found just by using scientific means. For the things "known" through scientific investigations are just a small amount of the deep reality inside the substance of everything. To put it another way, physics, to which most human beings tend to focus, is a limited aspect of the massive iceberg of spiritual things. Unfortunately, most human beings think that the spiritual world excludes physics, whereas, in fact, the physical world is just a tiny fraction of the surface of some visible aspects of the spiritual realm. The comprehensive description of the processes involved in the formation of the compounds (e.g., macromolecules, vitamins) found in living organisms and the diversity of organisms across the ecosystems would require hundreds of pages, therefore lengthening too much the size of the current book. Hence, I dealt with the origin of life in a different book.

As I was exploring laws which can explain life, I crafted the "law of species belonging," which I would like to briefly address here. According to that law, each living organism knows how to belong to its species and acts like the other members of that species. Wild animals or plants of the same species behave the same. According to that same law, nonliving things such as matter know how to behave like the other matter of the same kind. For instance, chemical elements of the same kind behave the same way under the same conditions and with the same internal composition. For instance, all hydrogen atoms of the same kind (e.g., same isotope) behave the same way.

However, human beings do not act the same. Because of their free will and the immensity and diversity of their imaginations and goals, human beings act according to what they want, provided no one is blocking them. Sometimes, they hurt each other and try to kill each other for diverse conflicting reasons. Wild animals of the same kind do not usually do this kind of wicked thing to each other. Some bulls can sometimes fight, but I am not aware of any wild animal that lays a snare for other individuals of that species just to catch, trap, hurt, or kill them. Nevertheless, human beings do that! Some human beings have even been known to eat human meat (cannibalism).

Birds do not act like fish, and vice versa. No mammal can decide to fly. However, human beings know how to take a drug or do something to themselves or to others to cause them to act like another species, even to the point of foolishness or madness. Human beings know how to elaborate laws that can catch and put their political opponents into jail or even kill them. Human beings know how to collaborate with evil spirits to travel in the air just to hurt others. Some use astral projections or broom trips to travel through the air. Although these remarks have nothing to do with the formation of human beings, they show that something is wrong with most of them and the sin within some of them that they do not want to address continues to take some of them down a path of destruction for their own lives and will even condemn their own souls if they do not stop what they are doing and give their lives to Jesus. Despite the effort of man to stop doing wrong, they cannot do it on their own. John 8:34-36 says, *"34 Yeshua answered them, 'Amen, amen I tell you, everyone who sins is a slave to sin. ' " 35 Now the slave does not remain in the household forever; the son abides forever. 36 So if the Son sets you free, you will be free indeed."* The good news is that we no longer have to remain in our sins. Romans 5:20 gives us hope and says, *"But where sin increased, grace overflowed even more-21 so that just as sin reigned in death, so also grace might reign through righteousness, to eternal life through Messiah Yeshua our Lord."* So how can you have that righteousness and eternal life? Romans 10:9-10 instructs us, *"9 For if you confess with your mouth that Yeshua is Lord and believe in your heart that God raised Him from the dead, you will be saved. 10 For with the heart it is believed for righteousness, and with the mouth it is confessed for salvation."*

One might ask what the date of sin's beginning is. Things were not like this in the beginning. For there was a time when human beings lived in perfect harmony with God and the world He created. I addressed that harmony and the birth of disorder later, but for now, let me say a few words concerning the duration of the stay of Adam and Eve in the Garden of Eden before their fall (sin). Indeed, although the Bible does not specify how long Adam and Eve lived before sinning, other Jewish literature suggests that they could have remained in the Garden of Eden for about 5.5 hours before being expelled on the same day they were created. Other references point to at least 7 years. Other literature suggests that those dates may even have been intentionally corrupted by some people, whose names I won't list here—if you want to know, contact me. Nevertheless, I used to think that 5.5 hours to 7 years was a very short amount of time until, in January 2021, I realized that, in the days before Adam and Eve sinned, 5.5 hours was a lot of time. For that stay occurred before the introduction of sins, and the human ability was much higher than that afterward, meaning that things that could take years to be accomplished today could have been accomplished in a matter of seconds. In the spiritual realm, spiritual beings are not limited by time and space as human beings. In other words, the short period of time that Adam and Eve could have passed in the Garden of Eden is like an eternity which was temporarily suspended or cut short so that the physical time could affect human beings until the fulfillment of the covenant that God had with them and which is connected to the passing of about

6000-7000 years before eternity resumes again! For instance, in the blink of an eye, an angel can accomplish things that could take a mere human years to achieve. For instance, I have heard testimonies about angels healing incurable diseases and accomplishing great miracles instantaneously. In other words, although the time in the Garden of Eden is similar to ours, human beings' abilities there have changed significantly. Creation happened in 6 days of 24 hours each, of course, and the passage of time has not changed since Adam and Eve were formed, but living things and the world they lived in have changed! To learn more about my perspective on the formation of living things, consult my books *"Turbulent Origin of Life"* and *"Origin of the Spiritual World."*

'Science180 Academy' Success Strategy
HOW TO RAISE RATIONAL CHILDREN IN OUR MODERN WORLD

In our modern secular world, and with the many things that kids are taught at school and over which parents have little control once the kids head to public school, parents have a lot to worry about. But it does not have to be that way. Universe-origin and life-origin scientist Dr. Nathanael-Israel Israel has discovered that, more than ever, parents have a crucial responsibility to rationally prepare their kids to have a strong worldview that properly embraces both science and faith so their kids are not pulled on one side by the secular education and on the other side by religious belief. But how can parents and their children achieve that common goal?

Listen to this Beninese-American scientist and mathematician Dr. Nathanael-Israel Israel to figure it out. Nathanael-Israel is the author of the acclaimed book *"How Baby Universe was Born"*, an easy to understand scientific book primarily written for children ages 7-12 years old to help them properly crack the code of the formation of the universe in a language they completely enjoy and that prepares them to fight any secular or religious theory that may try to rationally drift them away from the reality of everything! Sample questions that will get answered include the following and many more:

- How can parents use the latest breakthrough about the universe's origin to rationally raise their kids?
- How can parents prepare their children to be victims of the danger of wrong theories and dogmas on the origin of life and the universe?
- What can parents do to shield their children from the influence of religious and scientific beliefs that try to enslave them in the name of reason or faith?
- Why is wrong science not the only danger of raising rational children, but wrong belief as well?
- How can we help children to positively navigate the intersection of science and faith? Learn more at Science180.com/children

Science180: The Premier Organization That Scientifically Decoded the Creation of the Universe, Life, and Chemicals Accurately

CHAPTER 21

IS MODERN SCIENCE PREVENTING YOU FROM KNOWING THE INITIAL ORDER IN THE UNIVERSE AND THE EVENTS SURROUNDING THE END OF THE WORLD, OR WILL THE WORLD NEVER END BECAUSE IT IS GETTING MORE ORGANIZED? HMMMM!

Before finalizing this book that you are reading, I have written more than 9 versions of it, and in each, I edited the content to "try" to align it with the "needs" of the audience I am trying to reach, without leaving out many who also need to know the truth. Although I was freer to discuss the Bible in this version of my books on the origin of the universe, I must confess that I was not free to address all the aspects I would have liked to share with the Christian community. In fact, in the original version of this book, I devoted a lot of pages to the following chapters that I ended up trimming from this book:

- Perfect order in the universe at the end of creation
- Birth of disorder on Earth
- Current state of the universe
- Future technological revolution: the rise of evil technologies
- Rapture: Jesus transporting believers to heaven at His Second Coming
- Great tribulation: the apex of the Satanic reign
- Millennium Kingdom
- Judgment of the unbelievers
- Hell: the eternal place of punishment of unbelievers
- Paradise: the eternal dwelling of believers

CHAPTER 21: ORDER IN THE UNIVERSE AND EVENTS SURROUNDING THE END OF THE WORLD

I understood very well that many controversies surround those future events, and I also have my viewpoint in light of my understanding of the Bible and the origin of the universe. I initially devoted more than 50 pages to the above concepts, but due to space limits and also the controversy surrounding some of these concepts, and considering my main goal of using this book to reach out to as many people as possible by explaining to them how God created the universe (instead of dwelling too much on doctrinal things), I decided to remove these chapters from this book (for they contain things which may seem controversial for some readers) but to reserve them for other books, including my books on the end of the world and the age of the universe. If you join Science180 Academy of Creationism (see www.Science180Academy.com), you will also learn more about those mysteries. Nevertheless, let me say a few words about some of these concepts.

Indeed, although some people think that God created or formed things in nature only during the 6 days of creation narrated in Genesis 1, I need to emphasize that, based on my investigation, the formation of certain things may have continued even after the 6th day of creation. Although by the end of the 6th day of creation, some things in the universe might still have been undergoing changes that would lead to the formation of other bodies, the things God intended to create on Earth were finished. The Bible did not provide much detail about what happened to the precursors of other bodies after the 6-day story recounted in Genesis. In other words, after the 6 days of creation, the precursors of some bodies in the universe could still have been undergoing intense turbulence that shaped them. The misunderstanding of this fact caused some people to think that all celestial bodies were created by the end of the 4th day, yet the mathematics I did to show the perfect match between science and the Biblical account of creation suggested to me that many celestial bodies were formed after the end of the 6th day of creation, but all their precursors were formed before the end of the 6th day of creation. The precursors of bodies were like babies, which, after birth, began to undergo changes until complete development. Even up until today, some systems can still be going through turbulence like what birthed some bodies by the end of the 6 days of creation. In *"Turbulent Origin of the Universe,"* I provided ample details.

At the end of the 6 days of creation, God formed a perfect world, and He looked at everything and said it was good, implying everything was in perfect order. Being a God of order, God would not have stopped creating if there was any disorder in the universe He created. There was harmony between all particles, systems of composite particles, and the bodies they formed. Everything God created was for a reason. Looking at some miracles, I understood that God did not stop creating after the 6th day, but He is still creating even today.

Everything God created has a mission. Among the living organisms, there was order. From the living to the nonliving things, everything in the universe has a purpose. Each living thing knows how to behave so its life does not hinder that of others. Wild animals can communicate with human beings. I guess we shouldn't be surprised that even in the Bible, God used a donkey to tell Balaam that an angel was in front of him when Balaam was pressing his donkey to move forward, while the donkey, seeing the angel, had stopped.

Although the world was created perfect by God, it did not take long for some of its inhabitants to begin damaging it. In the scientific community, the term "entropy" refers to disorder, while, according to biblical evidence, any actions that contradict God's commandments can sow disorder in the world. The first reported action that launched the entropy in the world was the rebellion of Satan and the angels who followed him. It was after that revolt that Satan incarnated himself as the serpent to tempt Adam and Eve in the Garden of Eden. I devoted pages to that in my book on the end of the world.

Since the fall of Adam and Eve, the Earth has been a victim of many sins committed by living things. In the end, the current state of the universe is a chaotic dominance of sin, while a few living things obey God's laws. Because of disobedience, some creatures have been violating God's laws, which should maintain all creatures in perfect harmony; the portion of the universe they can influence has been degrading. While some degradations are visible, others are spiritual and invisible, yet their impact extends beyond the current world. More details on the current state of the universe can be seen in my book on the end of the world.

Knowing the impact that my findings on the origin of the universe will have on science, I think that the publication of all of the versions (scientific, popular, prophetic, biological, chemical, pseudepigraphic, children's versions, etc.) I did for this book will unleash a wave of discoveries which will lead to an unprecedented scientific revolution filled with breakthroughs, innovations, technologies, and inventions of various kinds! Therefore, I would like to warn those who will use my findings to ensure they honor God and serve others, and to work on their salvation so they do not seek the temporary pleasures of life on Earth, only to face eternal regret in hell.

For some reason, I feel like difficult times are ahead for the world and particularly for believers, after which eternal celebration will come for those who believe in Jesus. As more knowledge is poured into the world, technology will advance, but unfortunately, because some of those who control the world are very wicked, many evil things may be created or invented and then, unfortunately, used for malicious purposes. In the end, many of those who believe in God will be persecuted on this Earth before their time to rule eternally comes. Hence, God is also raising up a generation of giants who are being anointed for powerful tasks worthy of the end times we are living in. I am part of this generation of anointed and powerful believers who will shake the world in its last days before the rapture, which is when Jesus will take home those who believed in Him and have followed His instructions. I also provided more detail on the future technological revolution in my book on the end of the universe.

Although Christians have various interpretations of the rapture (the concept according to which some believe that Jesus will be transporting believers to heaven at His Second Coming very soon), most Christians believe that God will one day judge the world, sending some people to hell and others to heaven. As to the date of those events, no human being knows, but I did some mathematics which proves that we are very close to the end. Around that date, many Christians believe that there will be what is called the great tribulation, which is the apex of the Satanic reign, followed by the Millennium. Kingdom (a 1000 years reign), and the end of the world, the beginning of the stay in hell (the eternal place of punishment of unbelievers),) and the stay in paradise (paradise: the eternal dwelling of believers). I recognize not everyone agrees with these concepts, and many interpret them differently. For those who want to know my views on them (in light of my investigation), I've written books on them. If you would like to know more about the content of this chapter, including my calculation of the age and end of the universe, please visit www.Science180.com/books.

Another Book by Nathanael-Israel Israel:
ORIGIN OF THE SPIRITUAL WORLD

ONLY ONE ANCIENT BLUEPRINT HAS THE RELIABLE POWER TO HELP YOU TO ACCURATELY DECRYPT THE SPIRITUAL ORIGIN AND HISTORY OF EVERYTHING IN THE UNIVERSE

Countless books talk about the origin of the universe and of life, but this amazing book is the first and the only one that has undeniably explained how the formation of the universe and everything in it was truly revealed in the rejected and hidden scriptures such as the Books of Enoch and others. In *"Origin of the Spiritual World,"* you will:

- Discover deep-rejected secrets that have prevented humankind from unearthing the beginning of the universe
- Plainly see the scientific proof (hidden in scriptures) of the formation of the Earth, the Moon, and the Sun in a matter of days, a historic revelation that bizarrely and shockingly matches the scientific data as scientifically proved in *"From Science to Bible's Conclusions,"* a popular book written by Dr. Nathanael-Israel Israel
- Properly use the lost and rejected scriptures to articulate the process by which the universe was formed, and use that insight to improve your understanding of the Bible, innovate in your domain of interest, and improve your life perpetually

- Empower and align yourself with the historic breakthrough that has done what no other discovery has ever done: accurately unlock and decode mysteries concerning the origin of the cosmos and its content using scientific keys revealed in ancient scriptures that some elites have concealed (Science180.com/pseudepigraphic)

- Discover and apprehend the complex formation of the universe and life without leaving out the challenging questions that people of all ages have been struggling to answer for thousands of years, while the answers were hidden

- Find more joy in life through a clear interpretation of old and fresh revelations about the creation of the universe astonishingly backed by modern science, which some people wrongly think opposes the Bible

- Make a difference and blaze new trails for those who depend on your leadership

If you believe in God but have some origin-related questions whose answers you cannot find anywhere, not even in the Bible, and if you want to tap into historically neglected revelations to answer fundamental universe and life questions, then be sure to get a copy of *"Origin of the Spiritual World"* today.

Dr. Nathanael-Israel Israel happens to be the discoverer of the historic mathematical equations that scientifically demonstrated that the Earth was formed 2.82 days, the Moon 3.32 days, and the Sun 3.69 days after the beginning of the universe, therefore confirming the Biblical account of creation that revealed about 3500 years ago that the formation of the Earth was completed on the 3rd day, while that of the Moon and the Sun was completed on the 4th day of creation. Nathanael-Israel Israel is referred to as the "Undisputable Specialist of all Questions at the Intersection of Science and Biblical Creation". Learn more about this rare scientist at Israel120.com.

CHAPTER 22

IS GOD HIDING SOME UNIVERSE-ORIGIN SECRETS BECAUSE SOME CHRISTIANS ARE REFUSING TO BELIEVE IN THE GOD THEY WANT THE NONBELIEVERS TO ACCEPT–ARE THEY REALLY DOING THAT?

If you think major prophets, televangelists, and other ministers do not or cannot make huge mistakes about the Bible while on the pulpit, think again—we all do (to some extent?), and some of those lies will surprise you! What?! Buckle up and listen to what is coming up next!!

God hid some secrets and prevented some human beings and angels from discovering them quickly, but He revealed some to His people, the Israelites being the first.

- Why does God reveal secrets to those who believe in Him?
- Why and how did God first reveal Himself to the Israelites?
- Why did God not detail the creation story and future events?
- Why are great scientific discoveries made despite God hiding some secrets?
- Why must certain secrets be sealed until the end, when knowledge will increase in the world?

Those are questions I detailed in the initial version of this book, but due to space constraints and relevance, I will reveal mysteries about the last two points on the list above. Let me start with this question: Why are great scientific discoveries made despite God hiding some secrets?

Indeed, God, who said, "*Seek, and you will find*," cannot prevent people from finding what they seek, although He is not always the giver of the "solution." Some people seek life and find it from the giver of life (Jesus), while others seek death and find it away from God or without knowing God. Likewise, God is the source of some good discoveries, while the devil and human imagination produce negative things.

On December 5, 2013, I perceived that, after sin entered the world, God did not break or destroy the laws He used to create the universe. Otherwise, the universe would have collapsed, and natural laws of creation would be undiscoverable. Instead, God punished the creatures that failed and left most of the relationship between the created things intact. Some of those relationships are what scientists have discovered and then forged to fit their prejudicial beliefs. One of the main things that was successfully broken is men's communion with God. In other words, sin disrupted the direct communion that once existed between God and human beings. As orphans or widows, the corrupted spirits, souls, and bodies of human beings started selfishly searching for the so-called "solution" to their problems, while ignoring the real problem. Therefore, when Satan helped human beings to invent science, they began searching for knowledge by questioning God's creation instead of listening to and obeying God Himself. They progressively built scientific methods to "learn" from nature.

Because of the revelation that their ancestors received from God, the Jews have contributed to science (particularly physics, medicine, and biology) more than any other culture in the world. For example, people of Jewish descent have won more Nobel Prizes than any other culture on Earth. This does not mean that all Jews obey God, but regardless of their walk with God, they were exposed to some knowledge early on, which the other nations did not have the chance to know until today or until recently. A careful analysis of the gospel, particularly John 6-17, suggests that most Jews and some of their friends and allies were not and are still not able to properly interpret the intent of the law that God gave them. The history of scientific advancements also showed that scientists who believed in God made greater breakthroughs than unbelievers. Some unbelievers also discovered many things, but their unbelief destroyed their own lives and those of others.

Science could have experienced many more breakthroughs if Christians had obeyed God more and sought more of His direction in all their scientific efforts. Unfortunately, several Christians believe in evolutionism and many other theories that deny God, therefore compromising and sandwiching themselves between creationism and many philosophies of unbelief. Consequently, these Christians or "so-called" Christians cannot receive deep revelations from God because God cannot compromise by showing them things which they will credit to Satan or humankind. In other words, if we obey God and are willing to glorify Him with our deeds, He will do great things through us. Hence, the imagination of unbelievers is limited. If God had not limited human ability and potential, the unbelievers would have already wrecked the creation. Nevertheless, God has allowed some scientists to discover a few secrets that were originally hidden from previous generations. Because all intents of human beings and even of some unbelievers are not always evil, a residual or minimal quality of goodness is found in some inventions, which, in fact, though not 100% pure, express the limit put upon science in order to avoid a detrimental catastrophe. After all, a good intention of an unbeliever can never produce as much good as God's power working through a believer in Christ.

Despite God hiding many secrets of creation, human scientists have been discovering some things. The failure of certain scientific studies can be God restricting evil or God answering the saints' prayers, restraining the Antichrist's takeover. As progress is made, more damage is usually inflicted on the world, although the apparent results or fabrications seem to show the opposite, or are forced to show the devastating impact. Some human inventions negatively affect nature and its constituents more than the profit some people get from them. Hence, science has been paving the way to transform this Earth into an inhabitable planet. For instance, some electronic products (which are so loved by human beings nowadays) are causing health problems that their inventors and consumers would never acknowledge.

To prevent human beings from reaching and damaging other places in the universe, God also designed natural laws to make it difficult and costly to leave Earth and cross other celestial bodies without pain, delay, and a high price. Hence, it is expensive to escape Earth and go into space. Likewise, if God had made it easier for human beings to travel at the speed of light, they could have already damaged the universe and even tried to reach and fight God Himself if they could reach Him. Therefore, being very smart, God formed human beings with limited material, which is subject to weaknesses and environmental constraints, thereby breaking or slowing down some evil human imagination. Even if human beings try to use science to reach other celestial bodies at all costs, their own bodies will limit them. For they can die if they dare not consider their physiological constraints. God is very smart!

God did not intend for man to understand everything that He has done. This partially explains why He forbade Adam and Eve from eating the fruit of the tree of the knowledge of good and evil, which was meant to open their eyes to certain things. God also knew that His creatures are limited and it is impossible for them to fully understand creation. That is also why His grace grants us access to things we would never have deserved on our own. Therefore, let's be careful with science, for without a divine insight into the spiritual world and the unseen, some scientists will never explain the universe, but will be distanced from God forever.

At this point, I will elaborate on why and how certain secrets must be sealed until the end, when knowledge will increase in the world. Indeed, the Bible predicted that knowledge will increase at the end of the world (Daniel 12:4). As Daniel was trying hard to understand some of the prophecies given to him, God told him that some of them are secrets that must be sealed until the end (Daniel 12:8-9). As I came across this verse on April 18, 2014, I knew I needed to be very careful about how I handled what I was discovering about the formation of the universe. Although in those days I did not know I would be writing thousands of pages on the secrets of the world's formation, I was very reserved about sharing anything until the right time. Therefore, instead of trying to understand everything, I urge you to realize that it is more beneficial for human beings to act within the limits of the knowledge revealed to them or that they have discovered, which does not go against God's word but confirms it. For I have come to realize that when people force themselves to understand things beyond their limits or spectra, they can sin by believing or proposing wrong theories. Hence, there are many things I discovered that I could not even share with the whole world yet. But the day is coming when every hidden thing, particularly those hidden by men, will be made known: *"For there is nothing hidden that will not be revealed, nor anything kept secret except that it would come to light"* (Mark 4:22).

Knowing this, I felt more tolerant toward outstanding men of God (including those who some people mistakenly think always hear from God all the time), when they make dangerous mistakes in their interpretation of the creation narrative and other things in the Bible. Because I did not want to embarrass any of them, or to cause people to start criticizing them for their mistakes, I will not name any of them here, but I would like to address some of their thoughts so the readers may know some of the possible contemporary heresies supported by some "famous" preachers today.

Besides our Lord Jesus Christ, Moses, who wrote the laws, and Enoch, who revealed deep mysteries in his books, I don't know of any other person who has sought out and unearthed the scientific secrets about creation as I did. Yet, I want to humble myself and give God the praise and glory, for He alone knows it all. I used to believe everything some prophets said without questioning them (as they taught us), but after listening to their teachings on the creation of the world and other things, I was very disappointed by the heretical beliefs of some of them. This experience with creation opened my eyes to be more careful and not always to swallow every teaching from anyone (including the prophets), but to always confront them with the Bible, just as the Bereans (see the Bible's Book of Acts) did in the days of the apostles. Nevertheless, I still believe in the prophetic and respect prophets, but I have just come to realize that most of them do not know as much as they claim. Others support wrong things, thinking they are always hearing from God. As we respect and honor every man of God, including the prophets, let's be very careful not to start viewing anyone as a major human being on Earth who knows everything! Only God is omniscient, and according to what I discovered, God has not and will never allow any human being to know everything or be able to do everything, as some claim! Respect men of God, but do not let anyone fool you into giving away or giving them your God-given right and power of using your mind to think through some of what you believe! God will never condemn you for using your mind, which He gave you to think, question things, and seek His help for answers instead of putting your confidence in men who can deceive you; for after all, regardless of the level of his or her anointing, every human being is fallible.

When unbelievers make mistakes about their unbelief, believers are usually quick to attack them without providing scientific proofs that can satisfy their needs, while some believers also make significant mistakes in their interpretations of the Bible. Even among believers, it is important that people educate themselves on scientific facts so they can be more effective and helpful to the unbelievers they seek to evangelize. Otherwise, believers cannot reach the unsaved if they keep supporting heretic teachings that oppose the Bible as well as science. I hope to share my findings with some of those men of God, including where I think they are wrong, and that they will seek the truth and correct their mistakes for the glory of God. But will they listen? Nevertheless, I believe in God and his prophets. For no man is perfect, and everybody is going to make some mistakes in life, even famous or major or senior prophets saying wrong things that God did not tell them, for many things are meant to be sealed until the end. And those who ignore this fact can even reject the true words of God spoken by some of those prophets who fail to properly interpret the creation narrative and other things. Likewise, many of them are not always speaking the words of God as they claim. In the end, only God, who knows everything, will judge one day. Meanwhile, let's be tolerant and seek God more than the knowledge of things that are not properly to be decrypted before their appointed times. Therefore, I thank and praise God for allowing me to interpret part of the creation story. To God be all of the glory!

Another Book by Nathanael-Israel Israel:
TURBULENT ORIGIN OF THE UNIVERSE

THE FIRST AND ONLY SCIENTIFIC BOOK THAT ACCURATELY EXPLAINS EVERYTHING YOU NEED TO UNCONVENTIONALLY, EASILY, AFFORDABLY, AND ENJOYABLY DECODE THE UNIVERSE'S FORMATION

In "*Turbulent Origin of the Universe*", filled with great diagrams and digestible scientific facts, you will discover, learn, or get:

- The all-in-one, proven & uncomplicated scientific formula that accurately decoded the formation of the universe, and that explained the birthdate of the stars, planets, satellites, asteroids, and all other celestial bodies in the universe, so you can position yourself to stay on top of your competitors and avoid repeating crucial mistakes that many people have ignorantly made at their own perils

- Extraordinary, unprecedented, accurate insights into the first factors (e.g. early universe physics) that defined the history and formation of the universe so you can tap into deep scientific secrets you ignore, and set yourself apart from others

- The new physics that will revolutionize science forever and land you into a zone of original ideas that improve lives nonstop regardless of your expertise

- The 4 simple things without which it is impossible for anyone to ever understand the formation of the universe, think accurately, work differently, achieve, or perform better for superior results
- The verified key to move the cosmological mountains of misunderstanding, so you can confidently free your mind from doubts, improve your health, and prevent you from any danger connected with sticking with wrong assumptions
- Save time and money, and enjoy your life once you remove errors holding your true understanding of the universe's origin captive
- Historic scientific proof of whether a planet was formed in 2.82 days, whether a satellite was formed in 3.32 days, and whether a star was formed in 3.69 days after the beginning of the universe; so you can creatively produce and address a broader work spectrum by learning how to effectively communicate with and establish unusual connections between otherwise disconnected and disparate scientific data
- Why the scientific community has failed to sufficiently explain the origin of the universe and understand how existing theories have missed and undefined central ideas, and imposed limits on the vision of scientists
- Specific in-depth knowledge, up-to-the-minute information, and ideas so you can expand your market, cut useless costs, stop wasting time on inadequate projects, and start focusing on the profitable solutions (Science180.com/scientific)
- How Science180 Academy can strategically enlighten you and guide you to navigate and filter the massive data collected on the universe, so you can answer the world's most challenging questions, remove any scientific and philosophical cataracts that may be blocking you, and bring you many steps closer to your best life
- How to better resonate with your target market that is craving something original that breaks wrong explanations of the universe's origin

Get *"Turbulent Origin of the Universe"* today to begin an incredible journey of accurately decoding the universe and change your life forever!

Dr. Nathanael-Israel Israel is told by people that he is the #1 universe-origin, life-origin, and chemical-origin expert. He is the founder of Science180 and the author of many books on the origin of the universe and its content. To learn more about how he may help you, visit Israel120.com.

CHAPTER 23

WHERE DOES GOD DWELL, AND WHY IS HE FIRE AND LIGHT?

Because God is the creator of the universe, the proper understanding of the universe requires a foremost belief in God Himself. Hence, I cannot explain the origin of the universe without addressing who God is. In the first draft of this book, this chapter on God was the first I wrote after the introductory chapter, but because some people may want to see the proofs of the scientific demonstration of creation before learning about the Creator, I moved it to this section of the book. In this chapter, I do not intend to review all the revealed attributes of God, but I will focus on a few that align with the story I told in this book and can provide insight into the formation of the universe. In other words, because no human being can fully know everything about God, I chose to focus on a few revelations, knowing that many other great books address God's nature, characteristics, and other aspects.

23.1. God, the uncreated Creator, has no beginning and no end

The Bible declares that God has no beginning and no end. He is the Alpha and the Omega. His dwelling has always existed and is different from the universe He created. In other words, before creating the universe, God was living somewhere that should be as old as God Himself. In fact, it can be inappropriate to talk about God's age, for His age supposes a beginning that no human being can fully apprehend. It is not by chance that God is also called the "Ancient of Days," for He existed before anything called "day" today existed. Therefore, when we see God someday, we may better understand his existence and how He came into being from nothing, if that ever happened.

Anyway, before creating anything, God already declared when and how the end would come, creating a physical world from the spiritual one. Even so, there was a spiritual world that God also created. In other words, the spiritual things that God used to create the physical things have not always existed, but were also created at one point. There is nothing in the universe that was not created by God or whose precursor was not created by Him. I used the term "precursor" here because the world is nowadays filled with things not originally created or formed by God but made by human beings and others.

23.2. No man has ever seen God except Jesus

Except Jesus, who is God and who, during His stay on Earth, became incarnate (John 1:1-14), no human being has ever seen God (1 John 4:12). All the other people in the Bible who claimed or thought they had seen God had only encountered an angel of God. Even Moses, who claimed to have seen God, and Jacob (also called Israel), who claimed to have fought with God, did not really encounter God but rather an angel sent by God. While on the mountain top receiving the 10 commandments, yet having to cover his face in the presence of God, Moses encountered God through the burning bush. Just as Jesus also said, no one has ever ascended to heaven to see God (John 3:13). Nevertheless, it is important to mention that some patriarchs, such as Enoch, may have met God. For instance, the Bible declares that Enoch pleased God so much that God took him. The Book of Enoch also describes Enoch's meeting with God and his journey. I detailed this story in *"Origin of the Spiritual World."*

Some prophets (even contemporary ones) have claimed to have met God. Likewise, I have heard from people who testified that they met Jesus either in person or in a dream. Because I believe in the supernatural, I cannot discuss those claims. However, up to the time I am writing this book, I have never met God face to face, but as I was finishing the writing of this book on the origin of the universe in 2020, I dreamed twice about Jesus, ... and on one occasion, He dressed me in a white cloth ... Based on what the Bible says. God lives in believers today, for believers in Jesus are the temple of the Holy Spirit.

Although God has a head, ears, eyes, nose, back, feet, arms, and many other organs found with human beings, His nature and characteristics go beyond what the human brain can grasp. Jesus said that He knows the Father, and He who has seen Him has seen the Father.

23.3. Hidden message behind the fire of God

Fire is one of the characteristics or natures of God that caught my attention, and that goes well with the formation and composition of things in the universe. As I explained later in this book, it is not by chance that almost everything in the universe contains some form of fire in it. Even cold things contain energy that can be released as fire. For the most part, when precious materials are mentioned, people think of those on Earth or in the universe, such as gold, silver, and diamonds. Yet, God is not made of gold, silver, or diamonds, or anything most people would likely, first and foremost, treasure on this Earth. For instance, although some people may not perceive fire as the most precious thing on Earth, God is fire, and the Bible is filled with references to that. According to other Jewish books, God is a burning fire from which sparks fly outward.

I never understood why God is light and fire until I discovered some of the characteristics of the initial matter in the universe, which I called "turbulent prima materia", and which I extensively addressed in *Turbulent Origin of the Universe*. Indeed, the original material that God created in the universe is like fire, not the exact fire that constitutes God, but something near that He fashioned in many ways so that He alone can be the supreme being, while His creatures are made of materials that He can control forever and ever and that He can transform as He wishes to allow some of them to interact with Him. Hence, to this day, no human or spiritual being can see, visit, or touch Him without God Himself allowing such an interaction, first transforming that person. That is why, before the believers see God face-to-face and touch him, they will be transformed into a different body compatible with the environment of God's dwelling. If God were not fire and light, He could not have created a world so filled with them. Because God never changed and will never change, He could not have changed His nature after creating the world. He has been as He has been even before the beginning of the creation.

As a metaphor for His nature, God has placed many rituals involving fire and light in the commandments He gave to the Israelites and other people He wished to interact with. He also manifested many times as fire. For instance, God appeared to Moses in a flame of fire: "*Then the angel of Adonai appeared to him in a flame of fire from within a bush. So, he looked and saw the bush burning with fire, yet it was not consumed*" (Exodus 3:2). In the days of Moses, God caused fire to rain on Earth (Exodus 9:23). God used fire to lead His people. For instance, as the Israelites were migrating from Egypt to the Promised Land, a pillar of fire led them by night, giving them light (Exodus 13:21), a sign of God's presence with them. Sometimes, God descended on Earth to speak to Moses in a pillar of fire as was the case on Mount Sinai: "*Adonai had descended upon it [Mount Sinai] in fire. The smoke ascended like the smoke of a furnace. The whole mountain quaked greatly*" (Exodus 19:18. Moses himself testified that "*the appearance of the glory of Adonai was like a consuming fire on the top of the mountain in the sight of the children of Israel*" (Exodus 24:17).

In the book of Leviticus and others, God told the Israelites that some of their sacrifices, offerings, and other rituals must be made with fire, therefore portraying the importance that God gives to fire, a component of God that He represents on Earth as a physical fire, which in fact also has a spiritual component.

After the Israelites built a tabernacle or temple for God, a fire would also rest on top of it. The consumption of the offering by fire and the ascension of the smoke of the sacrifices toward heaven were signs of God's acceptance of the corresponding rituals. Because the fire on the altar was to be kept burning continually and never go out, the priest had to burn wood on it each morning (Leviticus 6:5). All of these rituals symbolize God's continual desire to dwell with men. Those who presented the offerings of God made by fire were supposed to have no defect (Leviticus 21:21).

During those days, anyone who touched the offerings of Adonai made by fire became holy (Leviticus 6:11), signifying the impact of God's interaction with men on human holiness. Even animals must be roasted with fire before being eaten, and no animal was supposed to be eaten by boiling it, but by cooking it over fire. Those who played with that fire ritual died, including the sons of the high priest who used a strange, unauthorized fire on the altar (Leviticus 10:12). Fire was also used to apply some judgment, including burning some people until they died.

Moses went to bluntly mention that *"God is a consuming fire, a jealous God"* (Deuteronomy 4:24). Likewise, God devoured His enemies and the enemies of His people with that same devouring fire. Just as He can devour His enemies with fire, so also one day He will throw into the inextinguishable fire (Matthew 3:12) those who initially accepted Him and then reject Him or those who never accepted Him at all. On some occasions, the Israelites testified about hearing the voice of God speaking from the midst of the fire (Deuteronomy 4:33): *"From the heavens He made you hear His voice to instruct you, and on earth He caused you to see His great fire—you heard His words from the midst of the fire"* (Deuteronomy 4:36). Sometimes, the Israelites were afraid of the fire, for it was not ordinary (Deuteronomy 5:5). The 10 commandments that people across the globe refer to today were given to Moses on two tablets of stone written by the finger of God from the midst of fire (Deuteronomy 9:10).

Those who believe and obey God are not destroyed by His fire, but are purified by it, and even if they are put into a physical fire, they may not burn, but they will be delivered, as was the case of Daniel's 3 friends (Shadrach, Meshach, and Abednego) in the furnace set by Nebuchadnezzar (Daniel 3:26-27). When Jesus came in this world as a human being, part of his mission was to baptize people with fire (Matthew 3:11). During His ministry on Earth, Jesus affirmed that He came on Earth to pour fire on the Earth: "*I am come to send fire on the earth; and what will I, if it be already kindled?*" (Luke 12:49). Other translations render the verse as: *"I came to pour out fire on the earth, and how I wish it were already ablaze!"* (Luke 12:49). While those who reject God will spend eternity in an inextinguishable or unquenchable fire, which is hell (Matthew 3:12), those who believe in God will be enlightened by the light coming out of God. In the last days, God will throw the devil into the lake of fire and brimstone (Revelation 20:10). Furthermore, the last enemies which will be thrown into the lake of fire are death and Sheol (Revelation 20:14). Beforehand, the Earth will also be destroyed by fire as I explained in another book. It is as if everything God created started with a form of fire and will end in a form of fire but with 3 different destinations: heaven, hell, or the end of existence. Certain creatures that exist today will not go to heaven nor to hell, but they will stop existing at the end of the world.

As I close this segment on fire, I would like to pinpoint that the fire of God should be expected to be different than the conventional fire found on Earth, and which Wikipedia (2018) defined as: "the rapid oxidation of a material in the exothermic chemical process of combustion, releasing heat, light, and various reaction products". I personally think that no chemical elements should be found in God, for He existed before designing and forming the universe and its chemical elements. Because He never changed, He could not have changed His nature using some physical things He created later. Hence, fire and light are a perfect description of God, yet God is not made of the same fire and light we see on Earth or in the universe.

In short, God displayed a very high intelligence by forming things with a limited fire, yet He Himself is an unapproachable fire. Indeed, as explained throughout this book, everything God created contains energy and fire in various ways. God is very smart to use the "same" fire He is made of to form things in nature but to delimit them beforehand by compressing, spiraling, winding them, or making a pile of them in such a way that the energy of the creatures cannot be released completely without destroying or transforming them, and also that the creatures cannot change their location and still remain completely the same, or come into God's presence in their current form without being transformed or destroyed. Thus, everyone cannot dwell in the presence of God, but only the elect or chosen ones, to whom God will grant many privileges. Controversially, God could not visit the Earth in all His glory, else He would have to destroy the Earth just as a big stone cannot fall on an egg without breaking it. So, for God to come to Earth to redeem humankind, He had to abide by certain earthly rules, including being born of a woman. But when He comes back a second time on the last day, He will bring His glory!

In summary, I felt that God's intention for human beings was to work on their relationship with Him, but most people prefer to pursue the "fulfillment" of their desires, regardless of how those desires align with God's universal plan. For instance, while God did not provide much information about the origin of the universe, He gave human beings many instructions on how to be saved. In their turn, some human beings tend not to focus on how to be saved and on life after this life, but rather on remote things in the current universe that they cannot see holistically.

23.4. Encrypted message behind the light of God

Besides fire, light is another main characteristic of God. God is the Father of lights, with whom there is no variation or shifting shadow (James 1:17). In the following terms, Apostle John also reported that God is light: *"Now this is the message we have heard from Him and announce to you—that God is light and in Him there is no darkness at all"* (1 John 1:5). The light that describes God should be different from the light. He created during the formation of the universe, which consists of photons. Indeed, light was one of the early matters that God formed (Genesis 1:3): *"Then God said, 'Let there be light!' and there was light."* However, although this statement suggests that no reaction occurred when God called light into existence, other Jewish books provide more details on how light was formed. In *"Origin of the Spiritual World,"* I addressed those mysteries. I decided not to illustrate them here because some people may not believe in them and might stop reading this book. I am well aware that, just as some unbelievers decide to never read the Bible, some Christians and Jews have made up their minds to never believe in some lost and rejected books well quoted in the Bible that I handled in *"Origin of the Spiritual World".*

Just as there is a physical light, there is also a spiritual light. For instance, talking about the spiritual light, Jesus said: *"I am the light of the world. The one who follows Me will no longer walk in darkness but will have the light of life"* (John 8:12). Those who believe in Him will not remain in darkness (John 12:46). Referring to the scripture where Jesus said He is the light of the world, the light that Jesus referred to can also mean a physical light. For, in the New Jerusalem which will descend from heaven to the New Earth that God will create after destroying the current earth we live on, there will be no need for light, for God Himself will illuminate the city with His glory. In other words, Jesus is really a light. Indeed, at the end of the current age and by the beginning of eternity, after God throws the devil into the lake of fire, God Himself will sojourn among His people forever and ever, and His light will illuminate them: *"And the city [the New Jerusalem] has no need for the sun or the moon to shine on it, for the glory of God lights it up, and its lamp is the Lamb. The nations shall walk by its light, and the kings of the earth bring their glory into it"* (Revelation 21:23-24).

If I dare to report something I read in some Jewish writings, the nature of Adam before his fall was that of a bright light, and Adam and Eve never experienced darkness until they were ejected from the Garden of Eden. In other words, because of their bright nature, darkness could not sustain itself around them. Wherever they went, there was light coming out of them. They were like a lamp that could not be overpowered by darkness. This implies that if Adam and Eve had not sinned, they could have maintained their bright nature and could never have experienced darkness, although the Earth could still have days and nights, while wherever they go, light should emanate from them and enlighten their environment. I was sad to learn that Adam and Eve were shocked by their first night in darkness after their fall, for before they sinned, they had never known what night was. I am not mentioning this story of Adam and Eve here to back all of the lost or rejected scriptures, but to point out that there are ancient stories which support the existence of mysterious secrets surrounding light, darkness, and why and how human beings are apparently the most vulnerable to environmental conditions if the leisure and comfort they find in their clothes and houses can be removed and force them to live as wild animals. In other words, common sense should have been enough to know that, at some point in the past, human beings could have lost something of their original nature. Else, why, despite their greatest intelligence and ability as compared to all other wild animals, are human beings unable to live as wild animals? How come human beings cannot live naked without needing artificial clothes, regardless of the season, time of day, or environment? The loss of the original nature by Adam and Eve and their descendants also affected the protection against environmental conditions including cold, heat, etc., that human beings once had. Unfortunately, when people see wild animals rejoicing and playing even in the coldest places on Earth, such as the poles, they never imagine what happened to us, human beings, who cannot even sustain even temporary heat in the summer and cold in the winter! To learn more about those mysteries, please refer to *"Origin of the Spiritual World"*.

Although God is light, it is important to know that Satan (the enemy of God) can sometimes counterfeit himself as light and behave like an angel of light (2 Corinthians 11:14). Hence, discernment is required to judge things. Unfortunately, *"the god of this world has blinded the minds of the unbelieving, so they might not see the light of the Good News of the glory of Messiah, who is the image of God"* (2 Corinthians 4:4). The nature of light and darkness relates to many things in the universe.

23.5. Where was God before and during creation?

Aside from a few verses in the Book of Job, the Bible does not detail what God was doing before creation or where He was staying. Although many Jewish writings gave details missing in the Bible, I did not elaborate here on that, but I just summarized what can be known from the biblical account. In *"Origin of the Spiritual World,"* I reported what other Jewish books said about what God did before creation.

On February 21st, 2014, it appeared to me that human beings are limited in what they can know, while God is unlimited. Likewise, the universe is limited. Even if there were a demonstration that could explain the origin, birthdate, birthplace, and dwelling of God, the whole universe may not have enough space to contain books that could be written about such things. Worse, the human brain is limited in its ability to fully explain the infinite. The choice is given to those who do not want to believe in the power of the unlimited God to continue doing so at their own peril, which they will understand very late, at a time when the demonstration of the proofs about God and creation will no longer profit them anymore. In other words, now is the time to accept the proofs about God so that when the proofs are plainly given one day, the believers can benefit from the faith they placed in the Word of God. Another way of saying this is that the current age is not meant for any human being or any created-spirits to fully comprehend 100% who God is, what He was doing before creating the universe, and where He was living.

God, who can weigh the thoughts, heart, and kidneys (reins) of living things, is the only one who, in due time, can prove what He has done according to His wish. And you do not want to doubt Him now and eternally regret your current choice in the eternal pain that awaits the unbelievers, some of whom think that life will never end or end only on Earth.

Being light, God cannot remove His own light to accommodate the darkness of the night as seen on earth. Being a spirit, God's domain of existence cannot be limited by anything. Before creating the universe, He could have chosen to exercise His influence over any domain of space He wanted. Although some people believe that space was created with matter, I think that there is a type of space that is not the one containing the things that God created, which should have always been in existence since the ages when God was the only being in the entire universe and in His circumference. In other words, there was a space in which God was living and moving before He decided to create the universe. Because God is a spirit who can fill all things, His presence could have been felt throughout the space He dwelt in. But where was that space where God was living in then?

The Bible did not give details about where God dwelt before creation. However, the Bible says that God dwells in the heaven of heavens, suggesting that He lives in a place above all heavens, which can mean above all created things and systems of things or beings: "*God that made the world and all things therein, seeing that he is Lord of heaven and earth, dwelleth not in temples made with hands*" (Acts 17:24). Moreover, other Jewish writings mention where God was before creating the universe, how He was even moving in the world in which He existed before creation. To avoid offending some people, I decided not to share those stories here, but those who are curious can consult *"Origin of the Spiritual World"* for more details.

Those who believe that God the Father has never left His throne hold that He has never moved from where He has always been. Because God is omnipresent, the verbs "go," "come back," and "leave" do not apply to Him, for He does not need to move before accomplishing anything. Therefore, saying that God came to Earth and then went somewhere does not really account for God's complex nature. He can be anywhere and know anything at the same time without changing anything in Him or around Him. Human beings are limited by space, time, and some spiritual dimensions, but God is not. In other words, questions about what God was doing before creation make no sense in the spiritual realm and are merely human efforts, limited by the physical world, seeking to fully understand the incomprehensible before its time. For the day is coming when believers will understand many things hidden in the current age.

Some people also think that God could have been sitting on His throne when He was creating the universe. However, other Jewish scriptures (see *"Origin of the Spiritual World"*) explained that God made His throne using some of the initial matter He created on the first day of creation. Although some people can be offended at the thought that God has not always been on His throne but had to make it at one point in His existence, I have to clarify that the throne of God is not God Himself, and reporting that God had to create His throne at one point in history is not aberrant.

Anyways, the dwelling place of God is not on the Earth or in the created universe accessible to men and angels. Surrounded by a glorious cloud, light, and fire, God lives above everything, and nothing else exists above Him. According to my understanding of the scriptures and the scientific data, I think that the direction of the north pole points toward the dwelling of God. For reasons I cannot demonstrate here, I also believe that the distance separating the dwelling of God and the universe He created (at least that we inhabit) has increased due, for instance, to the expansion of the universe and also because I think the whole universe is descending away from the position it started. Another way of expressing this is that the universe is not only expanding, as science has demonstrated, but also descending from its beginning at a speed greater than that of any man-made spacecraft. This also suggests that, despite their progress in the exploration of the universe, human beings will never fully understand the beginning and the limits of the world unless they believe in God and His revelations, including some of the rejected and lost ones, which are filled with mysteries beyond the most sophisticated level of knowledge that science can ever understand before the end comes.

The universe accessible to human understanding is limited by a kind of membrane embedded in the vast distances separating celestial bodies, which will never allow human beings and their equipment to know everything. Yet, people who have counted their journey to heaven reported how quickly the travel was, thereby pointing to the ability of spirits to travel faster than human beings can imagine.

23.6. God the Father sent Yeshua (Jesus, God the Son) to create the world

In many epistles, Apostle Paul explained how Jesus created everything: *"For by Him [Jesus] were all things created, that are in heaven, and that are in earth, visible and invisible, whether they be thrones, or dominions, or principalities, or powers: all things were created by him, and for him"* (Colossians 1:16-17). The Book of Hebrews also says that Jesus, the son of God, made the whole world (Hebrews 1:1-2). During His ministry on Earth, Jesus often explained that He was sent by God the Father, whom He said is greater than He, Jesus the man (John 14-17). Therefore, I realized on October 31st, 2013, that it is impossible to understand the universe without grasping Jesus' mission during and after the universe's formation. Although Jesus lived on Earth as a man, He was the incarnation of God the Father and His image while on Earth. Although Jesus the man was not God the Father, Jesus the Son of God is believed to be God himself, and some people even say He is equal to God the Father. Nevertheless, according to the Book of John, Jesus himself explained on many occasions that He was not God the Father. There is no question that Jesus was God, but the real question was whether He was God the Father. Although Jesus Himself acknowledged that He is God, to my understanding, He never clearly said that He was God the Father. I understand that some people may use some writings in the

Book of John to justify that Jesus was God the Father, and I do not want to argue that, but I want to point out the difficulty that many people have in demonstrating, accepting, or reconciling the doctrine of the deity of Jesus and the Jewish belief according to which God is one and that there should not exist a God the Son different from God the Father! I did not seek to address this mystery in this book, for many messianic authors have properly handled it.

Whether Jesus, the Son of God, was God the Father or not, or whether people believed that He had to empty Himself to come down on Earth in the form of a human being to redeem it or not, He was the one who created the world. He came into the world about 2000 years ago to accomplish what the Adam was unable to accomplish about 3500 years earlier, living a holy life and allowing Himself to be crucified on the cross so that, through His suffering, the shedding of His blood, His resurrection, and ascension to heaven, believers can be saved by faith in Him, whom the Father sent. Therefore, He became the only way, the door to heaven for the eternal reunion of God and the human beings qualified to live in eternity with God, the creator.

The whole universe is the property of God, meaning it belongs to God (Psalm 24:1), and He can do whatever He pleases. God Himself told Job that everything under heaven belongs to Him [God] (Job 41:3). It is unfortunate that some people do not know or believe that the whole world belongs to God and that human beings are like temporary tenants on the Earth, which is one of the countless real estates that God has. Just as the owner of a real estate property can evict the tenants who do not live by the lease requirements, so also God will evict all human beings from this Earth one day and send some to hell and others to heaven. Those who do not know these mysteries reject God in all their deeds; some of them build very expensive mansions and adopt lavish lifestyles, not knowing that they are acting as tenants who are building their tents on the property of a powerful owner they refuse to respect and obey, but that they keep disobeying, insulting, and provoking until the day of their eviction comes. In other words, the day is coming when God will destroy this Earth to recreate a new one where He will dwell eternally with the believers, meaning the righteous descendants of Adam, while the unbelievers will spend eternity in hell. But the thing is that nobody knows the date or the hour the world will end: "*Heaven and earth will pass away, but My words will never pass away. But of that day and hour no one knows, not even the angels of heaven nor the Son, except the Father alone*" (Mat 24: 35, Tree of Life Version).

23.7. Why is God called God the Father and God the Son, but not God the Mother?

On December 31, 2013, I got some of the inspirations I addressed in this chapter. I also wondered why the Bible talks about God the Father and God the Son but never about God the Mother. How can there be a God the Son and God the Father but no God the Mother? In those days, it appeared to me that human beings were supposed to be the spouse of Jesus. The rapture (that I detailed in another book) that many believers eagerly await is described as a prelude to a marriage or a wedding ceremony between the Lord Jesus and the spiritual church, which consists of all believers, including the dead believers who will be raised from the dead and the living believers who will be transformed.

I used to wonder why Jesus (considered as God the Son) needs a bride, while God the Father does not need a wife so there could be a God the Mother. I later realized that God the Father did not and does not need a wife because He is perfect and needs nothing. In Genesis 1-2, Eve (the first woman) was given to Adam (the first man) because Adam was alone and needed a companion. Although God was alone before creating the universe (that's a topic I discussed in *"Origin of the Spiritual World"*), He had all He needed. Otherwise, He would not be God the Father if He had to create a woman for Himself. I therefore wondered why Jesus needed a wife in the form of the Church. Did Jesus have a need that God the Father wanted to satisfy by creating human beings? I realized that it could not have been so, for this would mean He was not God, who is perfect and needs nothing. I then realized that even before God created the world, God knew that human beings would later fall. Hence, before the foundation of the world, He planned a way to save them, and the incarnation of God Himself to come to the Earth to save human beings was a mysterious formula that is metaphorically hidden in sexual relationships done within a marriage. For God needed a formula to keep loving the world He created until He redeems the human beings He created in His own image and likeness. Otherwise, God would have destroyed His image, something He will never do, for He cares a lot for His name. Until today, the Jews are still scared to even mention the name of God. For those who called His name in vain were culpable of death! Therefore, before the beginning, Jesus Christ, known to the Jews as Yeshua, was put aside by God to save the world through a spiritual marriage. Consequently, the more believers spend time worshipping, praising, and obeying God, as well as focusing on heavenly things, the more intimate their relationship with God becomes. In contrast, the more the believers disobey God, the more salt they put into the wounds of Jesus, who was crucified for them.

I used to think that angels do not have a gender and that in heaven everybody will be asexual. But one day, I realized that the intimacy that believers will have with God is more than the sexual experience on Earth, which some married people view as the sweetest thing or one of the sweetest things, ignoring that the fleshly pleasure that some living things seek on Earth is an encrypted code of the joy that the believers will have in their eternal presence and communion with God, their maker, who did not initially intend for life to be so hard on Earth! In other words, to some extent, sexual "pleasures" are like a feel, taste, or primer of the real pleasure believers will experience in eternity, which is not a sexual pleasure as done or experienced on Earth. When God starts the celebration of His eternal marriage with the believers, the physical marriage will no longer apply or exist. In other words, by the beginning of the eternity in heaven, physical marriage as known today to living beings will no longer exist, for, by that time, earthly matter will be destroyed and replaced by an incorruptible matter, while the current body hosting the spiritual being of unbelievers will be replaced by an immortal matter that can last in the suffering of hell eternally or as long as God wishes.

The spiritual marriage that God will have with the believers is a more profound and secure communion than that which Adam and Eve had with God in the Garden of Eden, when they did not know many of the things most human beings know today. While evil people keep trying to possess the whole Earth and its inhabitants, they ignore the secret plan and code God has been running throughout history, which is leading the world toward a better end for the believers only, but a regretful surprise for the unbelievers. In other words, life and the pursuits people pursue during their lifetime are an encrypted code for the kind of existence they will have in eternity. In His majestic plan, God has defined a precise time He will evacuate all human beings from this Earth (either by their death or by their translation into another world) or by ending the world so that the exam or test of life can be judged by Him and retribution be given accordingly to each human being and spirit, including evil spirits that are already judged, yet many unbelievers ignorantly embrace things that make them feel good in this life, while they are unknowingly already dead spiritually though their flesh is still "alive".

As I finished this section on God the Father and the Son, I would like to recall that not only is God our Father, but He does not want us to call anyone else father on this earth": "*And call no man your father upon the earth, for one is your Father, which is in heaven*" (Matthew 23:9). There is only one Father (Mat 23:8-9) who is in heaven, but all believers are disciples of Jesus, brothers, and sisters. Jesus did not tell his disciples to call Him Father, but He always said: "*Our Father who art in heaven*" (John 14:17). About 2000 years ago, Jesus could have been perceived as the only son of God, but since Jesus died and rose from the dead, He is no longer the only son of God. Since then, many human beings who believe in Him have been given the power to become sons of God (John 1:12). In other words, Jesus is the first Son of God, the only one who rose from the dead and who is the spouse of all other sons of God. Although angels are very powerful, they are not sons of God. Angels do not have the same attributes as human beings, and angels will be judged by human beings.

Despite my efforts to strategically address certain attributes of God, I know many of the questions people have may remain. In the chapter on gravity, I addressed the encrypted message behind God's omnipresence. What shocks me is that, despite creating us and the world around us, God is not believed by most people, and those who went to school and made discoveries about nature did not honor God as they should have. As the names of the chemical elements suggest, most scientists who think they know the world are missing out and need to glorify God more.

As I wrap up this book, let me revisit its main question: How was the universe formed, and did God really form it? If so, how can we scientifically prove the existence of the God who created it? You have heard people say that it took millions or billions of years for the Sun, Earth, and Moon to form. But if you are a literal believer, you accept that God created the universe. But how did God do it? For about 3500 years, no human being has ever scientifically demonstrated with proofs that satisfied believers, skeptics, and other freethinkers who rejected the existence of a creator. In this historic book written for great people who are interested in knowing how God created the universe, I, Nathanael-Israel Israel — (the founder of Science180, the first and only organization in the world to accurately prove the formation of the universe from a perspective that satisfies both believers and unbelievers)—scientifically demonstrated that from the beginning of the world it took:

- 2.82 days for the Earth to be formed
- 3.32 days for the Moon to be formed and
- 3.69 days for the Sun to be formed

Let me also ask you:

- Is there a simple scientific recipe to solve the dangerous decline of the biblical worldview in Western nations? Is that even possible with the advancement of modern science?
- Why freethinkers and rationalists enjoy rejecting the Bible (the most rational story told before the birth of science) and how it affects their search for the truth
- Why have freethinkers and rationalists rejected the rationality of the Biblical account of creation too early, even before the scientific data needed to demonstrate it were collected?
- Why is accurately reconciling science and the Bible the greatest opportunity of our time?
- Can most Christian leaders refuse to take a stand on the 6 literal days of creation but expect atheists and freethinkers not to argue that God is simply unnecessary?
- Why in Science180 Creationism do smart people tired of existing creationist and anti-creationist theories see the accurate scientific decoder of Biblical creation?

I know you may be tempted to answer these questions yourselves, but avoid going down the wrong paths that have caused some people to lose faith in God. It is better to get the accurate answer from the know-how expert, Dr. Nathanael-Israel Israel, the author of this book, the standout expert who accurately decoded the universe's origin, including the scientific formula that forces science to bow to the Bible. Do the following to get some help:

- Register at Science180 Academy of Creationism (www.Science180Creationism.com)

- Interview Dr. Nathanael-Israel Israel on your TV, radio, or internet show; visit www.Israel120.com/interview to learn more

'Science180 Academy' Success Strategy:
SCIENCE180 INTERVIEW REPORT (AKA SCIENCE180 INTERNET-TV-RADIO INTERVIEW REPORT)

Science180 Interview Report is the newsletter to read for guests and unconventional show ideas at the intersection of science and faith. Indeed, many hot questions are still unanswered on the road leading to the correct understanding of the origin of the universe, of life, and of chemicals. But most people don't know where to find the accurate answers to those challenging questions. What if, with one simple call, you can accurately answer all of those questions? You need to get in touch with or interview Dr. Nathanael-Israel Israel on your show, radio, TV, podcast, and even website, or invite him for a live presentation at your organization if your audience can benefit from any of the following show, talk, speaking, or interview ideas:

- Can anyone scientifically solve the most-asked questions on the universe's creation after 3500 years of rational demands?
- Can we explain the formation of the universe through natural processes without invoking evolution and the Big Bang?
- Can you really be scientifically 100% sure and prove that God created the universe?
- Can you scientifically demonstrate the Biblical account of creation without mentioning the Bible?
- How to talk to evolutionists, Big Bang proponents, atheists, and all other freethinkers about the universe's formation and have them beg you to teach them more about God, the Creator?
- Is faith better than science?

- Is Science really at war with religion?
- Is there a way to present the perfect scientific proof of God's existence?
- Is your church or pastor making you doubt God or the Biblical creation?
- What are the 3 world-shaking truths about the separation of science and faith nobody ever told you?
- What are the 7 Biblical Genesis details that seem bogus and stupid but really do scientifically defend the Biblical account of creation?
- What is the one simple scientific formula that will make anyone pay attention to the Bible?
- What is the only scientific story that atheists and evolutionists hear and automatically believe in the existence of God? Period.
- Does the Bible scientifically teach anything about the universe's origin that most people, including Christians ignore?
- How can we fix the tremendously dangerous trend according to which more people are denying God at the profit of secular theories because they think that it is impossible for science and faith to meet?
- How does the Biblical account of creation help to defend the existence of God?
- How people, including some fervent Christians, come to believe lies about creation and what they can do to change them so atheists can enjoy God.
- How to find truth, joy, and accuracy at the intersection of the secularly and biblically divided worlds?
- How to raise rational children in our modern world?
- How to scientifically prove without talking about the Bible that God created the universe
- What needs to be done about the dangerous trend according to which many Christians are abandoning their faith at the profit of secular doctrines that deny creation because they have been disappointed in and victimized by their creationist theories?
- What needs to be done to fix the extremely dangerous trend that causes more and more people to embrace evolutionism and trash the Biblical account of creation as if the Bible lied about the 6 days of creation?

- Why are most nations (governments) wasting millions of dollars on universe-origin and life-origin research they don't need—or do they?
- Why arguments against secular science are NOT arguments for creation
- Why are Christians abandoning wrong creationist theories that compromise with Darwinism and Big Bang?
- Why do secular rationalists and freethinkers think that Christians are irrational?
- Why freethinkers and rationalists enjoyed rejecting the most rational story told before the birth of science and how it affects their search for the truth
- Why is it so challenging for people today to choose the right explanation of the universe's origin?
- Why is reconciling Science and the Bible accurately is the greatest opportunity of our time?
- Why you don't have to embrace evolution or deny God to scientifically prove that God created the universe in 6 literal days

I know you may be tempted to answer these questions by yourselves, but avoid landing yourself on wrong paths that caused some people to lose contact with reality, it is better to get the accurate answer from the know-how expert, Dr. Nathanael-Israel Israel, the author of many books on the origin of the universe, of life, and of chemicals, and the standout expert who accurately decoded the scientific formula that forces science to bow to the Bible. Invite Nathanael-Israel Israel to your organization asap to hear his mind blowing answers to these challenging questions. If you would like to register for the Science180 Interview Report so we can periodically send you show ideas and opportunities related to the origin of the universe, of life, and of chemical particles, please visit Science180Interviews.com for more details.

CHAPTER 24

SCIENCE180 MODEL OF THE UNIVERSE'S CREATION BY THE GOD "THEY" DON'T WANT YOU TO KNOW, AND THEY WON'T TEACH YOU AT ANY UNIVERSITY OR CHURCH IN THE WORLD

What is the scientific power of the Biblical account of creation that "they" don't want you to know and they won't teach you at any university or church in the world?

Many books have been written about the origin of the universe, of course. But if you have read this book from its beginning until this point, you would have noticed that, since the beginning of humankind until today, no book has ever systematically proven the perfect match between the scientific data and the Biblical account of creation like this book and others I wrote on the origin of the universe did. Truth be told, as I was writing this book, I had many mysterious experiences that shaped my perspective about how to approach creation. Some of them shocked me, while others made me wonder why people still do not believe in God. Instead of summarizing everything I said in this book about the origin of the universe, I will just pinpoint a few mysteries that stuck with me the most as I reflected on my journey to this point. As you go over them, I hope you will find the information useful.

In the books that I wrote on the origin of the universe, I examined the existing scientific data from a fresh angle, integrating turbulence, a complex phenomenon that has been puzzling experts in that field for more than five centuries. For the first time in history, I scientifically showed that the Earth was formed on the 3rd day, and the Moon and the Sun on the 4th day of creation, just like the Bible says. In other words, for the first time in history, this book and 8 others I wrote (see www.Israel120.com/books) reconciled the Biblical creation story and the scientific evidence. Using scientific data, I discovered the formula for calculating the birthdates of celestial bodies.

I showed that, before creating the universe, God designed everything. Then, He spoke, and the first matter in the universe was born and became the precursor of all the bodies in the universe. Then, God caused the bulk of the initial matter to move, begin breaking into pieces, and reorganize into clusters of matter, leading to a cascade of breakups. In other words, as the bulk of the initial matter began moving, it split and reorganized into clusters. I invented the term split-gathering to indicate how the precursors of bodies split and reorganized into daughter bodies. As these clusters moved, instability arose within them and grew into turbulence. In the process, the matters being molded reached a stage near that of a fluid. The fluids were organized into layers, which interacted with one another. The top layers squashed the bottom ones and moved faster than the bottom ones. In the end, the speed of the bodies formed from the fluids was affected by the position of their precursors in the stack of fluids. During the breakup cascade, many types of body precursors were formed on scales ranging from galactic to microscopic. During these breakups, precursors of bodies split-gathered into smaller precursors and so on until the smallest scales in which no breakup of precursor could appear anymore. For instance, the precursors of galaxies split-gathered into the precursors of stellar systems, which split-gathered into the precursors of stars and the precursors of the bodies orbiting them, which split-gathered into the precursors of asteroids and the precursors of planetary systems, which split-gathered into the precursors of planets and satellites, and inside all these precursors, precursors of chemical particles were formed according to the intensity of the turbulence that shaped the systems they belonged to. All the fluid layers of the precursors underwent turbulent processes before being gathered into their daughter bodies.

When the precursors of galaxies were forming across the universe, the precursor of the Milky Way Galaxy was born, then split and coalesced into the precursors of stellar systems. It was at that point that the precursor of the Solar System was born, and then was split into the precursor of the Sun and the precursor of the bodies orbiting the Sun. I showed that the precursor of the bodies orbiting the Sun escaped the precursor of the Sun at about the Sun's escape velocity (617.6 km/s) and was organized into fluid layers that moved and split according to their positions. As the precursor of the bodies orbiting the Sun continued its movement and traveled for about 149,600,000 km (i.e., the distance separating the Sun and the Earth also termed the semi major axis of the Earth), the precursor of the Earth-Moon system split from the rest of the precursor of the bodies orbiting the Sun and then split-gathered into the precursor of the Earth and the precursor of the Moon. The time which elapsed up to that point was about 67.286 hours (i.e., 149,600,000 km divided by 617.6 km/s). This timescale is what I call the Earth's semi-major axis timescale.

As the precursor of the Earth-Moon system was splitting, the precursor of the Moon escaped the precursor of the Earth at about 11.186 km/s (i.e., the escape velocity of the Earth) and then traveled for about 384,400 km (i.e. the semi major

268
Nathanael-Israel Israel: Acknowledged as the Undisputable Specialist of All Questions at the
Intersection of Science and Biblical Creation

axis of the Moon), meaning a 9.546-hour trip, before reaching a position where its fluid layers were ready to be gathered into a spherical Moon. These 9.546 hours are what I call the semi-major axis timescale of the Moon. By this time, the precursor of the Earth swirled and formed its length. Using the scientific fact, I proved that just before the precursors of the celestial bodies wrapped their fluid layers into celestial bodies, they were as long as the circumference of the bodies they would form. For that final-stage swirling to occur, the fluid layers of the precursors curled at about the orbital speed of their daughter bodies. Knowing the radius (R) of the celestial bodies, I calculated their circumference as $2\pi R = 2 \times 3.14 \times R$, and divided it by the orbital speed of the bodies to get the circumference timescale of the celestial bodies, which is the amount of time it could have taken for their precursors to wrap around once they traveled to and reached their orbit. Using that formula, I showed that within 22.42 minutes, the precursor of the Earth swirled at about 29.78 km/s (the Earth's orbital speed) to form the Earth, whose radius is 6,378.14 km. Likewise, after reaching the orbit of the Moon, the precursor of the Moon swirled for about 2.96 hours at about 1.02 km/s (i.e., the orbital speed of the Moon) to form the Moon, whose radius is 1,738.1 km. In other words, the circumference timescale of the Earth is 22.42 minutes, and that of the Moon is 2.96 hours. I showed that the duration of Earth's formation equals the Earth's semi-major axis timescale and its circumference timescale: 67.286 hours + 22.42 minutes = 67.659 hours = 2.819 days. In other words, the time it took for the Earth to be formed was equal to the time it took for the precursor of the bodies orbiting the Sun (which embedded the precursor of the Earth-Moon system) to escape the precursor of the Sun and reach the orbit of the Earth, and the time it took for the precursor of the Earth to curl and form the Earth after the precursor of the Earth-Moon system split into the precursor of the Earth and the precursor of the Moon.

Using the same logic, I proved that the duration of the formation of the Moon (with respect to the Earth) was equal to the time it took for the precursor of the Moon to escape the precursor of the Earth and reach the orbit of the Moon (9.546 hours), and the time it took for the precursor of the Moon to curl and form the Moon (2.96 hours) after it reached its orbit. But because, with respect to the beginning of the formation of the Solar System, it took 67.286 hours before the precursor of the Moon was formed (see details above), I deduced that, with respect to the beginning, the amount of time which passed before the Moon was formed was: 67.286 hours + 9.546 hours + 2.96 hours = 79.795 hours = 3.325 days.

Furthermore, I also showed that, because the fluid layers of the precursor of Mercury were on top of the fluid layers of the precursor of the bodies orbiting the Sun, the moment it arrived at about the position of Mercury, all the fluids of the bodies orbiting the Sun could have reached that position as well. Hence, I used the time it took for the fluid layers of Mercury's precursor to reach Mercury's position as the maximum time required for a body's precursor to escape or move completely away from the Sun. Because the distance separating Mercury and the Sun is 57,910,000 km (i.e., the semi-major axis of Mercury), and because the escape velocity of the Sun is 617.6 km/s, by dividing that distance by that speed, I showed that the time required for the precursor of the Sun to be formed was about 26.046 hours (57,910,000 km / 617.6 km/s). About that time was needed for the precursor of the bodies orbiting the Sun to escape the precursor of the Sun. The passage of this time was also about when the precursor of Mercury (the innermost body orbiting the Sun) split from the bulk of the precursor's fluid, and it was also about what was needed before the precursor of the Sun could be formed. In other words, I proved that from the beginning of the formation of the Solar System until the precursor of the Sun was formed, about 26.046 hours have elapsed. At the expiration of that time, the fluid layers of the precursor of the Sun were as long as the circumference of the Sun (2 x 3.14 x 696,000 km) and swirled at about the orbital speed of the Sun (19.4 km/s) to form the Sun, whose radius is 696,000 km. That is to say that within 62.584 hours after its formation, the precursor of the Sun swirled and formed the Sun. Because 26.046 hours passed before the precursor of the Sun was formed, and then 62.584 hours were needed for its fluid layers to swirl and form the Sun, the total amount of time needed for the Sun to form was: 26.046 hours + 62.584 = 88.63 hours = 3.693 days.

These scientific calculations proved that the Earth formed 2.819 days (2 days, 19 hours, 39 minutes, and 33 seconds) after the beginning. That time is about the 20th hour of the 3rd day. I demonstrated that the Moon formed 3.325 days (3 days, 7 hours, 47 minutes, and 41 seconds) after the beginning. That time is about the 8th hour of the 4th day. Finally, the Sun was formed 3.693 days (meaning 3 days and 16 hours and 37 minutes and 49 seconds) after the beginning; that is about the 17th hour of the 4th day. This timeline is exactly what the Bible says. Indeed, the Bible reveals that the Earth underwent processes and that the formation of the Earth was completed on the 3rd day of creation (Genesis 1:1-13). The Bible also specified that the formation of the Moon and the Sun was completed on the 4th day of creation (Genesis 1:14-19). Considering the events the Bible addressed on Day 1 and Day 2, I showed that it alluded to turbulence and fluid breakup. Because on Day 1 the Bible also explained that the Earth had no form, but on Day 3 it was fully formed. I described that the Earth mentioned in Genesis 1:1 (*In the beginning God created the heaven and the earth*) was not the planet Earth known today, but its precursor. Many Biblical verses also suggest that God took great care in designing and implementing creation. For instance, I showed that the movement of the Spirit of God over the waters, as mentioned in Genesis 1, imparted abilities to the precursors of the bodies, enabling them to become what they are today. Considering that God is almighty, I demonstrated how His movement over the waters caused mighty turbulence in the waters. Therefore, I showed that the turbulence in the precursors of the celestial bodies can be traced back to the movement that the Spirit of God imparted to the initial matter, which God created with the word of His mouth.

Using this evidence, I showed that the days mentioned in the Biblical story of creation (Genesis 1) are literal 24-hour days. I pointed out three fundamental mistakes that prevented and keep preventing people from decoding the Biblical

270
Nathanael-Israel Israel: Acknowledged as the Undisputable Specialist of All Questions at the
Intersection of Science and Biblical Creation

account of creation:

1. Because they failed to distinguish between the Earth's precursor and the "finished" or "complete" Earth, some people mistakenly thought that the first verse of Genesis 1 is about the "complete" or "finished" Earth and the world that God created.

2. Because they distrusted the chronological account of the story in Genesis 1, some people reorganized the story of Genesis 1 as they pleased, wrongly thinking that Moses was not smart enough to state things according to the order they were created or formed, or incorrectly thinking that God did not recount a chronological story to Moses;

3. Because they distrusted the timing God revealed to Moses for the events in each of the 6 days of creation, some people developed theories that the universe was created over millions of years rather than the 6 literal days emphasized in Genesis 1.

I explained, for instance, how the creation story in Genesis 1 encrypted many secrets related to fluid movement by the Spirit of God on the waters, induction of turbulence, fluid separation or fluid breakup, gathering together or collection of fluid layers, formation of the atmosphere and crust of celestial bodies, formation of stars and satellites, and establishment of the rotational movement. I also explained how the story of creation in the Book of Job alludes to the precise measurement of the foundation of the Earth by God, the tilting of the celestial bodies, the transition of some precursors of matter from a gas-like state to a liquid-like state to a solid-like state, and the acquisition of shape by the daughter bodies of those precursors. The story in the book of Job also indicates the inclination of the precursors of bodies, a progression in the change of state of the precursors as their matter was being converted into their current shape. I also pointed to the processes suggested in the gospel of John 1: the Word of God that created initial physical matter, which was molded into physical things and beings in nature, where a battle exists between darkness and light, not only in the physical world but also in the spiritual realms. Many verses in the book of Psalms reveal that God made the world using His hand, just as a potter makes things. Even in the epistles of the Apostle Paul to the Ephesians and the Colossians, references to creation appear, such as God being not in the heavens but above heaven, answering the question of where God was before He created the world. God lived in a realm beyond the physical world that he created and that we see. In his epistle, Apostle Peter also helped us understand that the initial state of the universe's precursor could have been fiery; hence, the Earth will be destroyed by fire, and everything in it will melt.

I also showed that God purposely formed the Earth to host human beings, who, in turn, have a purpose that, unfortunately, most of them did not fulfill. Yet, the disobedience by humans cannot block God's plan, which includes a day of judgment and retribution for human beings according to what they have done on Earth. I also explained that the bodies in the universe were formed and maintained by a program found in both living and nonliving things. The diversity in the expression or implementation of this program of formation and maintenance of bodies or matter in the universe is vast and complex; hence, scientists have a hard time unraveling and modeling it with modern mathematics. I realized that the program that embodies the fundamental model of the universe is hidden in the split-gathering of the bulk of the initial matter that God created.

As I tried to understand why human beings have been unable to scientifically explain the formation of the universe, I realized that some scientific hypotheses are declarations of faith or unbelief, while unbelievers are like blind people walking toward a big pit. Indeed, scientific hypotheses are imaginations that some scientists want to prove, yet many scientists "refuse" to believe in some religions. However, once someone believes in something, they may not require certain physical proofs to accept it. The environment they live in may not be large enough to conduct research that would allow them to collect data to explain their beliefs. Therefore, by itself, faith can contain the proof of the thing believed. This implies that faith in God is more than knowledge, for it is a proven knowledge that needs no proof. Consequently, certain acts of faith enable believers to know more than unbelievers do. In other words, unbelief significantly limits the field of knowledge and wisdom of unbelievers. However, this does not mean that every faith is useful. How can one know which type of faith is real and true? This is where religion can come in and make it harder to explain to people which religion is better. This challenge should not cause anyone to reject religion and assume that all religions or faiths are wrong or that all are good. Unfortunately, that is the mistake many scientists have made by rejecting religion and focusing only on physical things.

Furthermore, some unbelievers are like a blind person who refuses to be led by someone who can see. Other unbelievers are like blind people walking near a big, deep ditch, yet refusing to heed the advice of those who can see and help them avoid falling into it. Suddenly, they fall into the hole and cannot blame those who were warning them and whom they refused to heed. Just as those blinds are responsible for their life and choices, so human beings are also responsible for the choices they have been making in this life, where they may overlook things they are not paying attention to or refuse to see. However, just as those blind people can escape falling into the ditch by listening to those who can see, so also some unbelievers can escape the eternal punishment of hell if they can listen to the believers. For there are things that the believers see with their spiritual eyes that the unbelievers will never see due to their darkness: disbelief. In other words, just as it is imperative for blind people to be aware of their blindness so they can be more careful, so also it is crucial for unbelievers to be aware of their own spiritual blindness. Unfortunately, like sick people choosing to be treated by incompetent medical doctors, so many human beings like bringing their problems to people who cannot solve them. Many have also been deceived by the physical appearance of things. In brief, it is impossible to give life to someone who prefers or chooses death. Likewise, regardless of how many proofs are presented to them, certain people will never believe the truth. Therefore, those who know they have the truth also need to be careful not to force people to accept it. Some will reject it after listening; others will not even want to listen before making the wrong decision, while others will embrace and applaud it even if they have not listened to all of it. Others will try to sell the truth and profit from it, instead of letting it change their lives for a better eternity, for some greedy people would prefer to eliminate the truth-bearers so they can profit from it themselves by preventing others from accessing it freely, as it was given to them. Aware of these mysteries, I know that my job is not to force people to accept that God created the world in 6 days and rested on the 7th, but to present my findings and let people continue making their own choice, just as God gives them the choice to believe or not to believe. Although the proofs of creation are overwhelming, I know I am a witness and not the one who will convince or judge people, but God will!

By the time I finished this book and the others I wrote about the universe, I realized why God did not bother giving too many details about how He created the world. There are too many details, and they would require a lifetime of study to understand them. Worse, thousands of years ago, when science had not even been invented, proper terminology lacked the means to express the process I addressed in this book. Moreover, most Christians have not spent much time properly meditating and investigating what happened during the creation that Moses recounted.

The gradual discovery of things by human beings is part of God's plan, so everything is not unearthed in one day, destroyed, or spoiled quickly, leaving future generations without certain resources corresponding to their times. Therefore, God gave Moses a summary, which Moses recounted in Genesis 1, which most Jews and Christians have believed but were unable to mathematically demonstrate until I did so for the first time in history. Science can explain the Bible, and vice versa. I also proved that the problem with the erroneous scientific theories is not because of the scientific data, but because of the ways scientists choose to interpret that data. Because the scientific data are immense, because the currently available scientific tools are not sufficient to properly analyze them, and because many errors are made in collecting them, it is certain that even a very strong explanation of the data will still contain errors.

Therefore, because no explanation of the universe is better than the one which God, the Creator of all things, gave, I would like to bring to your attention that my books must be put beneath the authority of the word of God, including the Bible and other true revelations God has been giving to His people. In other words, no theory and none of my writing should ever be put above the authentic revealed word of God. I did my best to ensure that none of my writings contradicts the Bible. The readers of my writings must be vigilant and careful never to use them to contradict the Bible. If anyone finds any contradiction between my writings and the word of God, I encourage them to revisit their thoughts and my writings to ensure they do not go against God. With time, some scientific findings may help explain certain data I may not yet have properly addressed. As my goal was to use science to explain creation, it is obvious that, over time, some of my theories must be revisited or updated to align with future scientific advances in a way that never contradicts the Bible. In other words, a scientist who wants his science to eternally benefit him must always ensure that his interpretation never opposes the word of God or compromises it. Any compromise with the word of God is a departure from it; therefore, we must always reject theories that compromise with the truth. For the truth is eternal, absolute, and can never hold its authenticity and supremacy in the midst of compromise. Never let your inability to explain the truth cause you to force the truth to say what it is not saying.

Despite the rise of evil in the world, despite the increasing abundance of truth, let us keep winning souls for Jesus. As we do that, remember that the mission of the believers is not to prove wrong or crush the unbelievers, including the evolutionists, but to help them to know and believe in God, and that goal cannot be achieved or reached if those who know or claim to know the truth must be using it to chase the sinners away as a light chasing darkness away. As we win more souls for Jesus, let us be more gracious toward people who do not believe in the same thing as we do.

Considering all I said in this book and the others I wrote on the origin of the universe, I believe that the best thing a human being can do in this world is to know and believe in God, and to ensure that he will go to heaven. Life on Earth is so short compared to the eternity to come; hence, it is pointless for a human being to run after earthly pleasures and needs while losing sight of eternal treasures. Therefore, I urge you to go give your life to God today if you have not yet. All it takes is to:

- acknowledge God as the creator of the whole universe and the Savior of humankind,
- acknowledge the wrong ways you have lived since you were born,
- ask God for forgiveness and for His grace to help you live a life that pleases Him,
- get a Bible and start reading it to better know God,
- find a church or a messianic congregation near you where you can fellowship with other believers,
- ensure to always check the accuracy of the doctrine of that church or messianic congregation with regard to the word of God as expressed in the Bible,
- do not settle in any church, congregation, or religious gathering where the word of God is not preached according to what you find in the Bible,
- keep learning about God, serving Him, loving your neighbors, and growing in your holiness by observing His laws and obeying His commandments,
- in all you do, always remember God, thank Him for everything, and seek to please Him above everybody or anything else.

If you do all of the things mentioned above, there is nothing in this life that can prevent you from being saved and enjoying heaven with God when He returns. And in that day, we will know more about creation than we do today, and we will also have all our needs answered. Before then, keep walking by faith and know that by God's grace, nothing will be impossible for you! To God, the Creator, be the glory, the honor, and the power forever and ever. Amen!

I cannot finish this book without thanking you, the reader, and telling you how much I value your time and the attention you gave it. Reading this manuscript from the beginning until this point shows that you are really seeking to understand the real explanation of the creation of the universe. I am glad you stuck with the reading to this point, and I know you gained a fresh insight into the origin of the universe. To help you grasp other aspects I did not cover in this book, I recommend you read my other books on the origin of the universe, life, and chemical particles. Also, please share the news with those around you, for many people are eagerly seeking the truth revealed in this book in the wrong places. Again, thank you for your time. If this book was helpful to you, I need to know. If there is anything I can do to address any question you have, I need to know. You can contact me at www.Science180.com, the site where I will also be sharing updates about my books and initiatives I am involved in, which may help you. On that site, you will find materials that will help you to better understand other aspects I did not delve into in this book. To stay informed about my works and news, please subscribe to my newsletter. If you would like to invite me to give a talk, collaborate with me on any project related to this topic, or consult me for my expertise, please contact me as well.

Finally, I thank God for the grace He gave me to believe in Him, and for spending 12 years (2013-2025) full-time working on this book and others related to the formation of the universe. I thank Him for taking care of my family and me during the crucial time, which led to the discovery and writing of books on the origin of the universe. To God be the glory, the honor, and all of the praise!

Nathanael-Israel Israel: Acknowledged as the Undisputable Specialist of All Questions at the
Intersection of Science and Biblical Creation

NEXT STEPS OF THE JOURNEY

Get free resources on Science180.com

If you have finished reading this book and would like to learn more about my discoveries and how they can help you, you are at the right place. Indeed, I am really committed to helping you address any questions you may still have about the origin, function, and fate of the universe, and how you can partner with me to achieve greater results.

To get free resources that will help you understand other aspects of the universe's formation not covered in this book, visit Science180.com and my personal website, Israel120.com. On those sites, I will be sharing guides and strategies to get the most out of my initiatives. I will also be sharing my favorite references, tips, next-step readings, and other important things in the pipeline that will help you, regardless of your field of expertise, interests, or needs.

Subscribe to "Science180 Newsletter": The only accurate universe-origin, life-origin, and chemical-origin newsletter in the whole world!

Be a part of decoding the universe's origin, life's origin, and chemicals' origin! Get origin-related news, information, discoveries, updates, announcements, reviews, articles, educational materials, and opportunities, from a holistic perspective not available anywhere else, so you can participate in and enjoy decoding the origin, current state, and fate of the universe and its content. You will also receive priceless tips about how Nathanael-Israel thinks, what his secrets and initiatives are, what he has accomplished, and what he recommends. Without any delay, sign up for the Science180 Newsletter today at Science180.com/newsletter. It is free!

Speaking engagement

Since the beginning of humanity, only Nathanael-Israel Israel, and no one else, has offered a scientific explanation of the universe's formation that perfectly matches science and the Biblical account of creation.

Unlike all other creationist speakers, Nathanael-Israel Israel is known as the first person in the whole world that calculated the mathematical equations that scientifically demonstrated that the Earth was formed 2.82 days after the beginning of the universe, while the Moon and the Sun were formed 3.32 days and 3.69 days, respectively, after the beginning of the universe, therefore confirming the 3500-year-old Biblical account of creation according to which the formation of the Earth was completed on the 3rd day, according to which the Moon and the Sun were completed on the 4th day of creation.

Nathanael-Israel Israel is also the first person in history to scientifically demonstrate that each day in the Biblical account of creation was literally 24 hours, a milestone that accurately reconciled science and the Bible, and that overturned the myth according to which some people have thought that each day of creation was millions of years (a misunderstanding that caused many people to deny God, the Creator). Therefore, Nathanael-Israel Israel ushered in a new era for the proper understanding of the Biblical account of creation and its application to decode the universe and its content for the benefit of humankind. He has provided an undeniable reconciliation of science and the Biblical account of creation. He is known as the one who offered the most accurate explanation of the Biblical account of creation. To book Dr. Nathanael-Israel Israel for a speaking engagement, visit Science180.com/speaking.

When you hire Nathanael-Israel Israel, you will learn from the historical specialist in universe-origin questions what the proven formula is for demonstrating the formation of the Earth on the 3rd day of creation and the formation of the Moon and Sun on the 4th day.

In addition to writing groundbreaking books and engaging in other business endeavors, Nathanael-Israel Israel is a renowned speaker whom you can invite to speak at your organization.

Values that Dr. Nathanael-Israel Israel can add to your life include the following:

- Irrefutable scientific proofs of the existence of God that will save you time and launch you into a zone of unlimited opportunities
- Unquestionable scientific proofs of how God created the universe
- Accurate demonstration of the historic formula that reconciled science and the Bible
- Rare expertise and tips that will increase your abilities
- Usefulness that will advance your impact regardless of your field of expertise
- Understanding of the world that will sharpen your perspective

- Critical information that will positively change your life
- Experiences turned into insight that will motivate and guide you
- Enlightenment that will help people, including Christians, to start using their brains instead of just praying and expecting God to do everything for them

For speaking inquiries, including how to book Dr. Nathanael-Israel Israel to speak to your organization or at an event, visit Science180.com/speaking for more details.

As the standout scientific authority who accurately decoded the universe, Nathanael-Israel Israel has been helping countless people across the globe discover and understand the complex origin of the universe without overlooking the challenging questions that people of all ages have been struggling to answer for thousands of years! As the true go-to expert on the formation of the universe and life, Nathanael-Israel believes that, regardless of age, background, culture, religion, or profession, everyone deserves to understand how the universe and life were formed and how they can leverage that knowledge to improve lives nonstop. Therefore, his groundbreaking discoveries of the formation of the universe, life, and chemicals have been broken down into books tailored to scientists (including physicists, chemists, biologists, mathematicians), laypeople or the general public, believers, and freethinkers; philosophers; children, etc., therefore maximizing the benefits to humanity. These historic, internationally acclaimed origin books include:

- "Turbulent Origin of the Universe"
- "Reconciling Science and Creation Accurately"
- "Turbulent Origin of Chemical Particles"
- "From Science to Bible's Conclusions"
- "Turbulent Origin of Life"
- "Origin of the Spiritual World"
- "How Baby Universe Was Born"
- "How God Created Baby Universe"
- "Science180 Accurate Scientific Proof of God"

When you hire Nathanael-Israel Israel to speak at your organization, you will:
- get specific in-depth knowledge, up-to-the-minute information, ideas, and insights about the universe's origin, life's origin, and chemical origins so that you expand your market, cut useless costs, stop wasting time on inadequate projects, and start focusing on the profitable solutions
- get relevant universe-origin stories that are specific to your field of expertise
- learn from a cooperative, flexible, and an easy to work with expert who will respond to your universe formation needs and position you to stay on top of your competitors

- interact with a renowned expert who will not just lecture you, but will help you sort out your origin-related questions using strategies to tap into deep secrets you ignore
- listen to an experienced expert who discovered outstanding secrets about the origin of all there is
- learn authentic information not from someone who just reads you a PowerPoint, but from the true go-to expert (when it comes to critical cosmological problems) who will share with you both his mistakes and successes that will help you get much closer to the better life you want to live
- revolutionize every origin-related domain with your accurate understanding of the universe's origin
- scientifically learn how the Earth was formed on the 3rd day of creation
- logically learn how the Sun and the Moon were formed on the 4th day of creation
- hear Dr. Nathanael-Israel Israel's personal selection and teaching of key topics that will help you break the code of the universe's formation and functioning, and strategically enlighten you; guide you to navigate and filter the massive data collected on the universe and its content so you know how to answer the world's most challenging origin questions; remove any scientific and philosophical cataracts that may be blocking you; and help bring you many steps closer to your best life today and forever
- hear the greatest scientific and philosophic lessons of some top scientists, philosophers, thinkers, and public figures who have realized historic mistakes they made in life (concerning the origin of the universe, life, and chemicals), and that they corrected thanks to the discoveries of Nathanael-Israel Israel, who founded Science180, and who is acknowledged as the scientist that truly decrypted the universe-origin for the first time
- Get world key lessons successful people have learned in life, and how people can learn from their experiences to improve their lives instead of repeating their mistakes that many people still ignore at their own perils

To book Dr. Nathanael-Israel Israel for a speaking engagement, visit Science180.com/speaking.

How you can make money by joining the affiliate program to sell Nathanael-Israel Israel's books

Greetings,

Do you want to make easy money by selling the #1 universe-origin, life-origin, and

280

Nathanael-Israel Israel: Acknowledged as the Undisputable Specialist of All Questions at the Intersection of Science and Biblical Creation chemicals-origin books on your website, newsletter, and by mail? You can start making big money by helping sell Science180 Books, including this one, on your website and through your network. Indeed, by now, you know that I operate a website called Science180.com, specialized in helping people across the globe to scientifically decode and understand the formation of the universe, life, and chemicals.

Your contacts, site, blog, forum, podcast, and newsletter may be admired among my target audience. Some of my products and services may be of interest to your audience. My books are the first in history to scientifically demonstrate the match between science and Biblical creation in a way that satisfies both believers and nonbelievers, a historic achievement and discovery that is revolutionizing our view of the origin of the universe, life, and chemicals for the benefit of humankind.

Imagine you have a website where you can talk to people about my books and services, and get a great percentage of every purchase they make on my site? Imagine you send a certain link about my books to your friends or network, and when any of your contacts buy a copy of my books, you get a percentage or a certain amount of what they pay on my sites. Imagine you can email your friends to spread the good news about my books, and when anyone uses that link to buy them, I give you something. Well! This is what the affiliate program is about. Apply today or learn more about it at Science180.com/affiliate. Likewise, if you own a website, you can apply for Science180's affiliate program, and I will send you a specific affiliate link that you will place on your website and newsletter, and if people click on it to buy my books, they will be led to my page, and after they buy, I will pay you a certain amount, sharing the profit with you instead of just verbally saying thank you.

Would you be interested in reviewing some of my products and services to explore becoming an affiliate? We have a wonderful affiliate program, and commissions are paid quickly and accurately.

If you are satisfied with the quality of our products and services, I am convinced you will also be impressed by our affiliate program.

I look forward to hearing from you

Nathanael-Israel Israel, PhD

Collaborate or partner with Nathanael-Israel Israel

If you have any lawful idea, initiative, or suggestion for a genuine partnership with Dr. Nathanael-Israel Israel or Science180, please visit Science180.com/partner to inform us.

How to be trained or mentored by or have a one-on-one consultation with Dr. Nathanael-Israel Israel

Hire Nathanael-Israel Israel to train you or your organization in the best ways to conduct yourself and your organization to align your initiatives with the real understanding of the origin of the universe, of life, and of chemical particles in a way that you will not hear anywhere else. Nathanael-Israel Israel offers training through the "Science180 Academy" program. For training purposes, please visit Science180Academy.com.

Visit Nathanael-Israel Israel's personal website to get great resources for free that you won't find anywhere else

To stay in touch with, Dr. Nathanael-Israel Israel, and to get updates directly from him, please visit his website, Israel120.com, and sign up for his popular newsletter at Israel120.com/newsletter for free.

Ask for a review

If you are a book reviewer or a professional wanting to review this book or others written by Nathanael-Israel Israel, please contact us at Science180.com/AskForReview

Donate and support Nathanael-Israel Israel's efforts and initiatives

To help humankind accurately understand the real origin of the universe and its content, as I have done in the groundbreaking books I published after 12 years of sacrifice, I need your financial support. Please consider donating to me or to Science180 by visiting Israel120.com/donate or Science180.com/donate.

Your donation will be used to help me continue doing what I did to birth these books that you enjoyed and that you know will help many people across the globe. No amount of money is too small or too big. Whatever you can give, please give.

Quantity discounts: Purchase Science180 books, including this one, in bulk at a special discount

To purchase Science180 books, including this one, in bulk at a special discount for sales promotion, corporate gifts, fund-raising, or educational purposes, or to create special editions to specifications, contact specialsales@science180.com or visit Science180.com/discount.

Buy a copy of Nathanael-Israel Israel's books for your friends, family, or someone else

If this book has been a blessing to you, and we know it has, please consider getting another copy and giving it to a friend, a family member, or someone you think it may help or challenge. If you want to get many copies, we can even give you a discount; just contact us as we previously explained.

Recommend Nathanael-Israel Israel's books to your organization

Because I know this book has been a blessing to you, I ask that you recommend it, along with others I wrote, to your organization, class, workplace, church, school, network, or clubs. Recommending this book will help others to tap into the blessings and opportunities that my books will open for them.

Share Nathanael-Israel Israel's groundbreaking discovery with others

To improve more lives, please share the findings of Nathanael-Israel Israel's books with others, for many people out there still do not understand how the universe was formed, and sharing your experience of reading this book will help them. If you enjoy Nathanael-Israel Israel's books, please help other people find them by writing a book review on your blog or on online bookstores, or write it and share it with us. Likewise, share and mention this book on your social media platforms (e.g., Facebook, Twitter, YouTube, etc.).

Follow Nathanael-Israel Israel on social media

In our modern world, social media has become a major factor in how messages spread across the globe. To ensure more people hear about the good news revealed in my books, I need you to follow me and share my content on your social media and in your network. To know the full list of my social media accounts and follow me, please visit Science180.com/socialmedia.

Share your feedback, critiques, testimony, experience, adventures, story, or comment about this book with me

How have Nathanael-Israel Israel's books and services at Science180 improved your life? I would love to hear from you.

To better understand how I can help you next and encourage others, I need to capture your testimony or criticisms. Please visit the feedback page, science180.com/feedback, to tell me:

- how this book impacted you or will impact you
- what you like or dislike or disagree with
- what you think, wish, or dream that I need to work on next
- what you wish to see in this book, but that was absent
- what shocked you the most
- what got your heart pumping as you were reading this book
- what you found more insightful or thought-provoking
- what you want to do to be a part of my journey
- how my work changed your life or someone else's life

Message from the publisher of this book

Just like Nathanael-Israel Israel, you can publish your book(s) with us, too. To get started and see how we may help you, please visit Science180Publishing.com today.

To contact Nathanael-Israel Israel or Science180

For any suggestions or questions, please visit Science180.com/contact and Nathanael-Israel Israel's personal website, Israel120.com. Feel free to ask me any questions you have about the formation of the universe, life, and chemicals.

REFERENCES

Clarke Adam (1832). Commentary on Genesis 1. The Adam Clarke Commentary. Retrieved on October 5, 2017, from www.studylight.org/commentaries/acc/genesis-1.html.

Eggers Jens and Emmanuel Villermaux (2008). Physics of liquid jets. Rep. Prog. Phys., 71(3):036601 (79pp). doi:10.1088/0034-4885/71/3/036601.

Feynman Richard (2006). The Feynman Lectures on Physics, The Definitive Edition, California Institute of Technology, Pearson, Addison Wesley, Vol. 1, p. 7-9.

George W. K. (2013). Lectures in Turbulence for the 21st Century. Department of Aeronautics Imperial College of London, London, UK, and Department of Applied Mechanics - Chalmers University of Technology Gothenburg, Sweden. Retrieved on October 19, 2017, from www.turbulence-online.com.

Israel Nathanael-Israel (2025a). Turbulent Origin of the Universe. Science180, Augusta, USA 683 pages.

Israel Nathanael-Israel (2025b). From Science to Bible's Conclusions. Science180, Augusta, USA 170 pages.

Israel Nathanael-Israel (2025c). Reconciling Science and Creation Accurately. Science180, Augusta, USA 299 pages.

Israel Nathanael-Israel (2025d). Turbulent Origin of Chemical Particles. Science180, Augusta, USA 397 pages.

Israel Nathanael-Israel (2025e). Turbulent Origin of Life. Science180, Augusta, USA 370 pages.

Israel Nathanael-Israel (2025f). Origin of the Spiritual World. Science180, Augusta, USA 151 pages.

Israel Nathanael-Israel (2025g). How Baby Universe Was Born. Science180, Augusta, USA 130 pages.

Israel Nathanael-Israel (2025h). How God Created Baby Universe. Science180, Augusta, USA 224 pages.

Israel Nathanael-Israel (2025i). Science180 Accurate Scientific Proof of God. Science180, Augusta, USA 214 pages.

Kolmogorov A N (1941) Dokl. Akad. Nauk SSSR 31 99–101).

Kolmogorov A N (1962) J. Fluid Mech. 13 82.

Luther Martin (1904). Luther on the creation. A critical and devotional commentary on Genesis. Based on Dr. Henry Cole's translation from the original Latin. Revised and Edited by John Nicholas Lenker, VOL. I. Lutherans in All Land CO., Minneapolis, MN, USA.

Malvern (2016). Malvern Instruments White Paper - A Basic Introduction to Rheology, www.malvern.com, Worcestershire, UK, 20 pages.

Monin A S and Yaglom A M (1975). Statistical Fluid Mechanics: Mechanics of Turbulence vol 2 (Cambridge, MA: MIT Press.

NASA (2018). Planetary fact sheets. Fact sheets of the Sun, planets, satellites, rings and selected asteroids in the Solar System. Author/Curator: Dr. David R. Williams, NASA Goddard Space Flight Center, Greenbelt, MD, USA. Retrieved on November 19, 2018, from http://nssdc.gsfc.nasa.gov/planetary/factsheet/.

Petitjeans Philippe and Frédéric Bottausci (2020). Structures tourbillonnaires étirées : les filaments de vorticité. Laboratoire de Physique et de Mécanique des Milieux Hétérogènes (UMR CNRS 7636) Ecole Supérieure de Physique et de Chimie Industrielles 10, rue Vauquelin, 75005 Paris. 13 pages.

Wikipedia (2018). Fire. Retrieved on July 13, 2018, from https://en.wikipedia.org/wiki/Fire.

Wikipedia (2019). Fluid thread breakup. Retrieved on November 2019, from https://en.wikipedia.org/wiki/Fluid_thread_breakup.

Wikipedia (2020). State of matter. Retrieved on April, 4, 2020, from https://en.wikipedia.org/wiki/State_of_matter.

Yinon Bentor (2016). Chemical Elements.com. Retrieved on February 27, 2023, from http://www.chemicalelements.com.

INDEX

Science180: The Go-To Source for Valuable Universe-Creation Information

G

H

Science180: The Go-To Source for Valuable Universe-Creation Information

Nathanael-Israel Israel: Author of "How Baby Universe Was Born"

ABOUT THE AUTHOR

Dr. Nathanael-Israel Israel is the renowned universe-origin, life-origin, and chemicals-origin expert, a scientist, mathematician, and philosopher. He is the founder of Science180 (www.Science180.com), the only American organization that fulfills its mission by providing unconventional and holistic expertise, authoritative advice, influence, and breakthroughs on major issues concerning science and creationism through research, publication, training, education, promotion of excellent dialogue between science and faith, facilitation of the public's understanding of science, public engagement, and the dissemination of top secrets of the origin of the universe, life, and chemicals for the benefit of humanity perpetually, in a historic way that authentically values the truth & honors God, the Creator. As the historic discoverer of the **Universe Creation Formula™**, meaning the all-in-one proven & uncomplicated formula for great scientists, laypeople, believers, nonbelievers, and all freethinkers to accurately decode the origin of the universe, of life, and of chemicals, Dr. Nathanael-Israel Israel is known and celebrated across the globe as the first human being that scientifically reconciled science and the Biblical account of creation in a way that does not force smart people to check their rational mind or their unwavering faith at the door. He has discovered and mastered and can teach you the accurate formula to methodologically talk to anti-creationists, evolutionists, Big Bang proponents, atheists, skeptics, and other freethinkers about the universe's formation, and they will beg you to know more about God, the Creator, whom they mistakenly deny. In addition to his passion for helping people to scientifically and accurately decrypt the formation of the universe and life, Dr. Nathanael-Israel Israel is a sought-after speaker on topics related to the origin of the universe, life, and chemicals. The insights of this Beninese-American have been featured on many media outlets in the US and across the globe.

He is a member of the American Chemical Society, American Association for the Advancement of Science, American Society of Agricultural and Biological Engineers, American Society for Microbiology, American Society of Biochemistry and Molecular Biology, Ecological Society of America, American Society of Agronomy, Crop Science Society of America, and Soil Science Society of America. The authoritative books of this internationally acclaimed scientist include:

- From Science to Bible's Conclusions
- Turbulent Origin of the Universe
- Reconciling Science and Creation Accurately
- Turbulent Origin of Chemical Particles
- Origin of the Spiritual World
- Turbulent Origin of Life

- How Baby Universe Was Born
- How God Created Baby Universe
- Science180 Accurate Scientific Proof of God
- Mathematical Proof of God's Existence at the Intersection of Science and Faith

By getting these well-celebrated books, you, your family, and your organization will surely embark on an incredible journey to accurately decode the universe and change lives forever! Find out more about this great scientist and how he may help you or your organization at Israel120.com.

Nathanael-Israel Israel: Creator of Science180 Academy (www.Science180Academy.com)